Project Management Simplified

A STEP-BY-STEP PROCESS

Industrial Innovation Series

Series Editor

Adedeji B. Badiru

Air Force Institute of Technology (AFIT) – Dayton, Ohio

PUBLISHED TITLES

Carbon Footprint Analysis: Concepts, Methods, Implementation, and Case Studies, *Matthew John Franchetti & Defne Apul*

Cellular Manufacturing: Mitigating Risk and Uncertainty, *John X. Wang*

Communication for Continuous Improvement Projects, *Tina Agustiady*

Computational Economic Analysis for Engineering and Industry, *Adedeji B. Badiru & Olufemi A. Omitaomu*

Conveyors: Applications, Selection, and Integration, *Patrick M. McGuire*

Culture and Trust in Technology-Driven Organizations, *Frances Alston*

Global Engineering: Design, Decision Making, and Communication, *Carlos Acosta, V. Jorge Leon, Charles Conrad, & Cesar O. Malave*

Global Manufacturing Technology Transfer: Africa–USA Strategies, Adaptations, and Management, *Adedeji B. Badiru*

Guide to Environment Safety and Health Management: Developing, Implementing, and Maintaining a Continuous Improvement Program, *Frances Alston & Emily J. Millikin*

Handbook of Emergency Response: A Human Factors and Systems Engineering Approach, *Adedeji B. Badiru & LeeAnn Racz*

Handbook of Industrial Engineering Equations, Formulas, and Calculations, *Adedeji B. Badiru & Olufemi A. Omitaomu*

Handbook of Industrial and Systems Engineering, Second Edition, *Adedeji B. Badiru*

Handbook of Military Industrial Engineering, *Adedeji B. Badiru & Marlin U. Thomas*

Industrial Control Systems: Mathematical and Statistical Models and Techniques, *Adedeji B. Badiru, Oye Ibidapo-Obe, & Babatunde J. Ayeni*

Industrial Project Management: Concepts, Tools, and Techniques, *Adedeji B. Badiru, Abidemi Badiru, & Adetokunboh Badiru*

Inventory Management: Non-Classical Views, *Mohamad Y. Jaber*

Kansei Engineering—2-volume set

- Innovations of Kansei Engineering, *Mitsuo Nagamachi & Anitawati Mohd Lokman*
- Kansei/Affective Engineering, *Mitsuo Nagamachi*

Kansei Innovation: Practical Design Applications for Product and Service Development, *Mitsuo Nagamachi & Anitawati Mohd Lokman*

Knowledge Discovery from Sensor Data, *Auroop R. Ganguly, João Gama, Olufemi A. Omitaomu, Mohamed Medhat Gaber, & Ranga Raju Vatsavai*

Learning Curves: Theory, Models, and Applications, *Mohamad Y. Jaber*

Managing Projects as Investments: Earned Value to Business Value, *Stephen A. Devaux*

Modern Construction: Lean Project Delivery and Integrated Practices, *Lincoln Harding Forbes & Syed M. Ahmed*

Moving from Project Management to Project Leadership: A Practical Guide to Leading Groups, *R. Camper Bull*

Project Management: Systems, Principles, and Applications, *Adedeji B. Badiru*

PUBLISHED TITLES

Project Management for the Oil and Gas Industry: A World System Approach, *Adedeji B. Badiru & Samuel O. Osisanya*

Project Management Simplified: A Step-by-Step Process, *Barbara Karten*

Quality Management in Construction Projects, *Abdul Razzak Rumane*

Quality Tools for Managing Construction Projects, *Abdul Razzak Rumane*

Social Responsibility: Failure Mode Effects and Analysis, *Holly Alison Duckworth & Rosemond Ann Moore*

Statistical Techniques for Project Control, *Adedeji B. Badiru & Tina Agustiady*

STEP Project Management: Guide for Science, Technology, and Engineering Projects, *Adedeji B. Badiru*

Sustainability: Utilizing Lean Six Sigma Techniques, *Tina Agustiady & Adedeji B. Badiru*

Systems Thinking: Coping with 21st Century Problems, *John Turner Boardman & Brian J. Sauser*

Techonomics: The Theory of Industrial Evolution, *H. Lee Martin*

Total Productive Maintenance: Strategies and Implementation Guide, *Tina Agustiady & Elizabeth A. Cudney*

Total Project Control: A Practitioner's Guide to Managing Projects as Investments, Second Edition, *Stephen A. Devaux*

Triple C Model of Project Management: Communication, Cooperation, Coordination, *Adedeji B. Badiru*

FORTHCOMING TITLES

3D Printing Handbook: Product Development for the Defense Industry, *Adedeji B. Badiru & Vhance V. Valencia*

Company Success in Manufacturing Organizations: A Holistic Systems Approach, *Ana M. Ferreras & Lesia L. Crumpton-Young*

Design for Profitability: Guidelines to Cost Effectively Management the Development Process of Complex Products, *Salah Ahmed Mohamed Elmoselhy*

Essentials of Engineering Leadership and Innovation, *Pamela McCauley-Bush & Lesia L. Crumpton-Young*

Handbook of Construction Management: Scope, Schedule, and Cost Control, *Abdul Razzak Rumane*

Handbook of Measurements: Benchmarks for Systems Accuracy and Precision, *Adedeji B. Badiru & LeeAnn Racz*

Introduction to Industrial Engineering, Second Edition, *Avraham Shtub & Yuval Cohen*

Manufacturing and Enterprise: An Integrated Systems Approach, *Adedeji B. Badiru, Oye Ibidapo-Obe & Babatunde J. Ayeni*

Project Management for Research: Tools and Techniques for Science and Technology, *Adedeji B. Badiru, Vhance V. Valencia & Christina Rusnock*

A Six Sigma Approach to Sustainability: Continual Improvement for Social Responsibility, *Holly Allison Duckworth & Andrea Hoffmeier Zimmerman*

Work Design: A Systematic Approach, *Adedeji B. Badiru*

Project Management Simplified

A STEP-BY-STEP PROCESS

Barbara Karten

CRC Press
Taylor & Francis Group
Boca Raton London New York

CRC Press is an imprint of the
Taylor & Francis Group, an **informa** business

CRC Press
Taylor & Francis Group
6000 Broken Sound Parkway NW, Suite 300
Boca Raton, FL 33487-2742

Printed on acid-free paper
Version Date: 20150916

International Standard Book Number-13: 978-1-4987-2934-5 (Paperback)

Library of Congress Cataloging-in-Publication Data

Karten, Barbara, 1948- author.
Project management simplified : a step-by-step process / Barbara Karten.
pages cm. -- (Industrial innovation series)
Includes bibliographical references and index.
ISBN 978-1-4987-2934-5 (alk. paper)
1. Project management. I. Title.

HD69.P75K365 2016
658.4'04--dc23 2015033141

Visit the Taylor & Francis Web site at
http://www.taylorandfrancis.com

and the CRC Press Web site at
http://www.crcpress.com

Contents

List of Figures

List of Tables

Preface

Projects are like a box of chocolates; you never know what you are going to get. Sometimes projects are simple, almost common sense. Other times, the complexity is overwhelming. But even apparently simple projects often run late, over budget, and fail to deliver on promised capabilities.

The Chaos Reports by the Standish Group suggest that 50%+ of information technology projects in the business world are challenged; they do not satisfy the scope, schedule, and budgetary requirements. My anecdotal observations suggest that personal, nonbusiness projects have a similar failure rate.

Maybe common sense is not simple. Perhaps there is an astounding lack of common sense among people working on projects. Voltaire suggests that "Common sense is not so common." Predictable project performance requires more than common sense.

Project management is a set of processes and a framework that enhances the opportunity for project success. It is a form of insurance for projects or situations where project success is important.

There are a number of structured approaches that guide the successful completion of projects in business environments. The Project Management Institute (PMI) documents a framework in its Project Management Body of Knowledge (PMBoK) standard. Project Management Institute Education Foundation (PMIEF) applies the

PMBoK principles to nonprofit, educational, and nongovernment organizations. Prince2 is another project management methodology that originated from the UK government.

These processes apply to our personal projects. This book translates these processes and techniques such that nonproject managers can easily use these proven approaches in a nonbusiness context for their own projects. It removes technical jargon, the need for computer software and hardware, and complicated organizational environments, describing the essential project management processes in a simple, straightforward manner.

Many people are not project managers at work, but might benefit from the application of project management principles on their job. Many of us have personal projects that are challenged in that they are often completed late and over budget, and "not quite right" in their features, functions, and capabilities. This book is for these people.

Author

Barbara Karten, PMP, has over 20 years of management, project/product/program management, and consulting/training experience. Since 2003, she has been an independent consultant and trainer delivering project management and business systems analysis training, PMP (project management professional) exam preparation training, general project management courses for continuing education credits, PMI (Project Management Institute) Chapter-sponsored study groups, and project management courses both online and on ground for graduate and undergraduate programs at local universities.

Barbara began her career with a focus on the application of systems methodology to human service systems development and operations. Over time, her interest in organizational effectiveness developed into a focus on requirements and business processes, and into positions as a business systems analyst for technology companies. Building on this experience, Barbara assumed responsibility for the management of projects and programs, for software and hardware in high technology companies.

In addition to her work with nonprofit human service agencies and for profit technology companies, Barbara taught project management at local universities, including University of Phoenix, Worcester Polytechnic Institute, and Northeastern University. She developed and updated PMP curricula and course materials for academia and local project management training providers. She has presented at industry conferences as a panelist and a speaker.

Barbara is a past president of the Central Mass PMI Chapter and is the Central Mass PMI Chapter liaison to the PMIEF (Project Management Institute Education Foundation), which addresses the project management needs of not-for-profit organizations, educational organizations, and NGOs.

She has the Project Management Institute PMP certification, Six Sigma Green Belt certification, and Microsoft Technology Specialist certification for MS Project 2010 (MCTS). From Boston University, she has certificates in Project Management and Business Systems Analysis, a master's degree in Computer Information Systems (MCIS), a master's degree in Urban Affairs (MUA), and a BA in psychology.

Barbara advocates project management skills as core competencies for life, education, career, and personal projects. She believes that people with project management capabilities are competitively positioned to succeed.

1

Projects

The Big Picture

Are projects a problem? Do your projects cost too much, take too long, or are just not quite right? If so, this book is for you; it is a no nonsense approach to projects—no jargon, just the facts.

Projects are all about change. They are temporary and have a defined beginning and end. They produce something unique; it may be a product, a service, or a result. Projects may be small or large, short term or long term, low risk or high risk. Planting a garden, planning a wedding, or changing jobs are all examples of projects.

Well-defined processes for managing projects have proven effective at delivering project success. This book applies these project processes to managing change in our lives. It views project management as a life skill.

Project management is for everyone; unlike other books about project management, this book focuses on the following:

- Everyday, low technology projects
- Simple, easy-to-use techniques for managing scope, schedule, budget, and quality
- Easily understood, jargon-free explanations of effective project management techniques
- Noncorporate projects
- Nonbusiness-focused project management

The book describes an approach modeled on a project process used successfully in businesses, not-for-profit organizations, schools, and other organizations. The skills and techniques are not unique to businesses and organizations; they are life skills available to everyone. This book presents a simplified version of the complete project management methodologies used in businesses. For simpler, less risky projects, the project and project team may not need all the components used in the business environment.

This is a journey, an adventure. Remember as children, we played a game called "Connect the Dots"? Once we connected all the dots, we had a picture of an animal, a basket, or a vehicle. As we progress through this book, we connect the dots necessary to complete our personal projects. This is our journey from the dreams you imagine to their realization.

The book includes a sample project in the text and a case study in the appendices that further illustrate the concepts explained in the text. This book also challenges the reader to select a project and, working along with the book, be the project manager and develop a project plan for this project. The project manager, working with the customers and funders of the project, defines the project and, based on this project definition, identifies how long it will take and how much it will cost. This begins the picture of connecting the dots.

The project manager writes this information down and builds on the initial information as he progresses through this journey. If he doesn't write down the information as he creates and collects it, he will likely have conversations as follows:

- Do you remember what we agreed to?
- I don't remember it that way. I think that we said… .
- We can't take that long; we need to be finished by the end of the school year.
- I'm sure we agreed on gluten-free alternatives for our guests.
- We said we'd not spend more than $xxx.

To avoid this confusion, the project manager writes down the decisions and agreements as they happen. He documents the information in the project plan.

Once the project manager has a picture of the goal, he identifies processes, methods, tools, and templates that enable the project to reach the goal.

The project manager is really doing two separate things. He identifies the project goals and a plan to reach that goal. Maybe the goal is a thing (like a deck or pool for the yard) or a successful event (like a wedding, trip, or community event), or it might be simply a change (like moving from one place to another or changing jobs). The project is what the stakeholders do to cause the goal to happen. Broadly defined, a stakeholder is anyone positively or negatively affected by the project. We often call the goal "the product"—so we have a product of our project. Building a deck, having a successful event, or effectively changing the place of residence happens as a result of a process—and collectively these processes are our project.

Prior to project initiation, the person paying for the project (called the project sponsor) identifies the high level project and product details in the form of a project charter. This serves as a contract between the sponsor and the project manager; it is the basis for the details of the project.

This book includes information about the project core, project risks, a project toolkit, project context, and project execution.

1.1 The Project Core

Once the project manager has an agreement and sign off on high level project and product details in the project charter, he collects more information and elaborates on the initial project and product details. He talks to the project stakeholders and secures their perspective on the project and the product of the project. He needs to understand their requirements, the "features and functions" they expect, and any considerations with respect to schedule, budget, and quality. He writes down these details and secures appropriate stakeholder sign off on this information. This is the project core. The project core contains the interrelated components of scope, schedule, budget, and quality. If a project has more scope (features/functions) or needs greater quality, then it will likely take longer and/or cost more. In this way, the core components are like a puzzle as shown in Figure 1.1.

If scope gets bigger or smaller or changes shape, then the puzzle pieces no longer fit. The project manager makes changes to other jigsaw puzzle pieces to get everything in synch (Figure 1.2).

1.2 Scope Component

For the scope component, the project manager elaborates the details of the project; he defines "the what" of the project. For example, if the

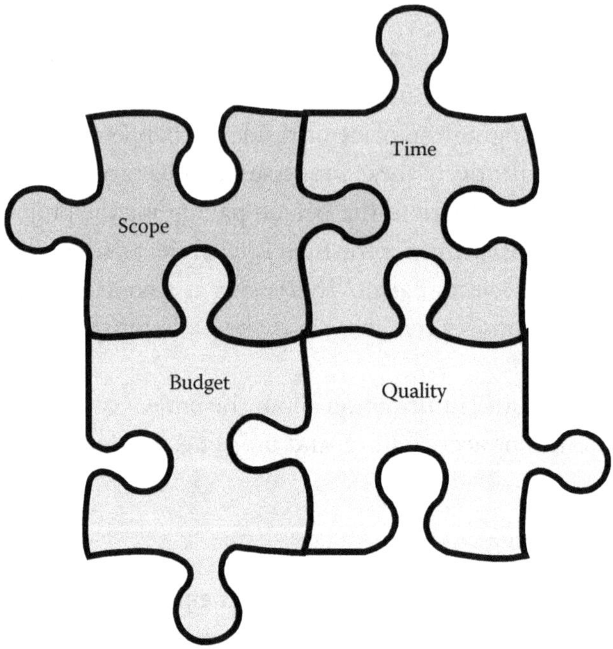

Figure 1.1 The project core puzzle.

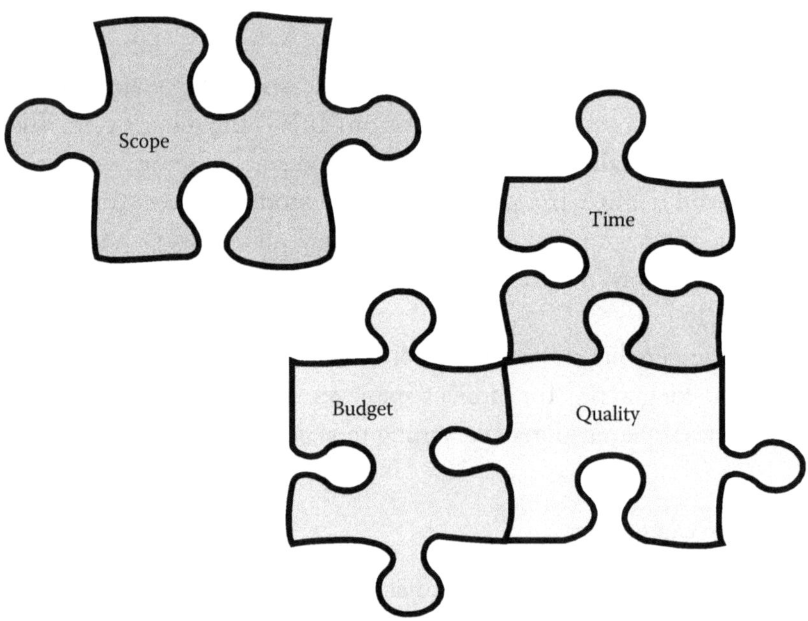

Figure 1.2 Project core puzzle mismatch.

project is to develop a game plan for a baseball team, he clarifies what is included in the effort such as

- The strategies for the players on the field
- The signals between various players
- Base stealing strategies
- The guidelines for players in the dugout
- The pre-game warm up
- The pre-season training camp
- The logistics for getting the team to the pre-season training camp and the details of the program
- Management of the press
- Ball park logistics including ticket sales, parking, food, and drink concessions
- And anything else included in the effort

The project scope determines the other core components. The level of effort, schedule, budget, and resources all vary based on answers to the questions above. Agreement on these details among stakeholders allows the project team to develop accurate details for effort, cost, schedule, budget, and resources.

1.3 Schedule Component

From a schedule perspective, is it enough to know that it takes 8 weeks to complete the project? In week 7, the project manager may realize that he will not complete the work per the original commitment, and now it's too late to fix it. If he schedules the work so that he knows what needs to be complete by the end of week 1, week 2, and so on, he identifies schedule slips early and may be able to take actions (add resources, do work concurrently) so that he still completes the work according to the original schedule completion date. The project manager's goal is predictability and completing work according to plan. He augments the end date of the project with intermediate schedule targets for key pieces of the work and documents this in the project plan.

A simple example of planning and managing a schedule might include attendance at a team meeting. A team member needs to be at a meeting at a local coffee shop at 8 a.m. He has to walk 8 blocks, and he knows each block takes 1 minute. Can he calculate a plan? Can he determine what time he needs to leave in order to be at the coffee shop on time? He may want to leave just a little bit earlier (put extra time in the schedule) just in case there is some delay en route? And once he has a plan, he can monitor against it. If it is 7:58 a.m. and he's 4 blocks away then he predicts that he won't be on time for the meeting. And, it's probably too late to hurry up (perhaps run a few blocks) and still be on time. If he discovers the problem at 7:55 a.m., he might be able to run the last couple of blocks and still arrive at the meeting on time.

1.4 Budget Component

There are similar considerations with respect to the budget. Is it enough to know that the team has $10,000 to complete the work of the project? Does the project manager need to know how much to spend for labor, supplies, permits, and the like? Does he need to know when the project will incur these expenses? If he doesn't have this information *and* track against it, he may find that he has spent the entire budget and still needs permits or additional work or materials. He documents this spending plan in the project plan.

For each package of work, the project manager identifies what it is, how much money (for resources, materials, services, permits) he needs, and how much time that he needs to complete the work package.

In later chapters, there are more details for scope, schedule, budget, and quality. The project manager balances this effort against the risk of running out of time, money, or finding that the project and product are incomplete, or, even worse, wrong.

1.5 Quality Component

Based on the project scope, schedule, and budget, the project manager considers quality concerns and understands what is "good enough." If the project is building a deck for the backyard, does the deck need to be perfectly level? Is a little bit of a slope ok? If furniture slides to one side or glasses slide off a table, there is probably too much slope and the deck lacks quality.

Quality is the extent to which the product and the project meet the needs of the customer. Perhaps the project is to build a pool in the backyard. If the customer (most likely your family) doesn't specify the size or whether it should be an inground or aboveground pool, then either would be a "quality" result. Unfortunately, customers often have expectations and needs that they don't specify; as a result, these expectations are implicit rather than explicit and are often not addressed. Even though either pool would theoretically be a quality product, in fact, the customer may not be satisfied, so quality is lacking. Similarly, a quality project meets the schedule and budget limitations set by the stakeholders. The project manager ensures that unstated needs are specified so that the product of the project is a quality pool that meets the customers' expectations.

1.6 Risk Component

The project core is the team's assessment of the project given what they know during planning; it is a projection into the future, much like a weather forecast. Even the best plan developed by the most experienced team with the best "crystal ball" has risks that the projections may not be 100% accurate. To manage this uncertainty, the team identifies risks that surround the project core (Figure 1.3). For the

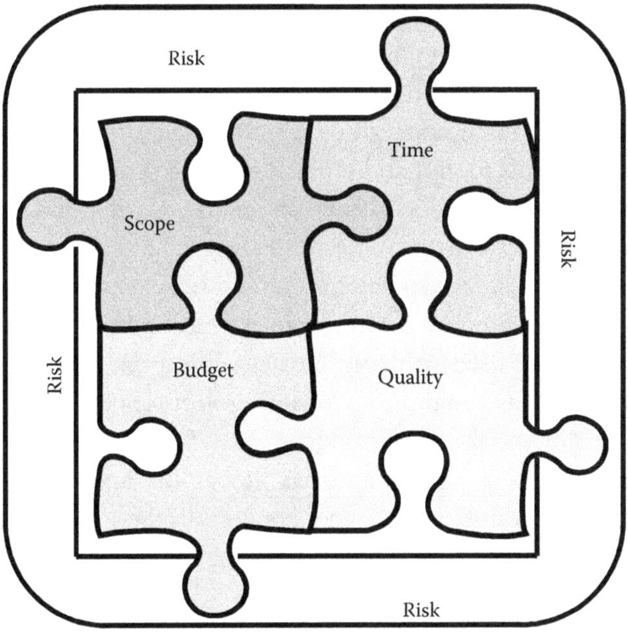

Figure 1.3 Project core puzzle surrounded by risk.

most significant risks, the project manager identifies risk responses and includes them in the appropriate project core component.

Once the team identifies the relevant risks, it determines what to do about them. Is it enough to recognize that a risk may occur, or should the team be proactive and ensure that threats to the project don't occur and ensure that opportunities do occur? The team takes advantage of opportunities and minimizes threats.

Managing risk on a project is like buying insurance. People have insurance for cars, homes, and health. Some people prefer to bear the risk themselves and self-insure. This might take the form of high deductibles, forgoing optional coverages and paying many expenses out of pocket. These are approaches that the project manager may use to respond to risks when they occur in the project environment. The project team assesses possible risks, and creates risk response plans for the most significant risks.

Once the project team has outlined the project core and the associated risks, the team decides how to accomplish the work.

1.7 Toolkit

When tackling new projects, it's more efficient to use what has worked in the past as the starting point. The team has a toolkit of documents, standards, tactics, and processes that have worked in similar situations in the past. The team adds guidance and suggestions from project stakeholders to this toolkit and may also look to project best practices to further enhance the documents, standards, tactics, and processes available for project work.

The toolkit includes documents, cross-project standards, tactics, and cross-project processes. Cross-project means that the standards and processes are used in many projects. The project plan includes documents, such as the project charter, project scope statement, and project schedule, among other documents. Cross-project standards might include risk categories and risk probability and impact scales. Tactics include brainstorming and graphic displays of information. Cross-project processes might include change management and decision making, among others.

These are repeatable processes. The toolkit facilitates the planning and management of projects. Repeatable processes mean that project stakeholders don't need to relearn processes, procedures, templates for each project. The work of project management gets easier as project managers gain experience with projects and stakeholders.

1.8 Project Context

The core project includes scope, schedule, budget, and quality. Risk is a key component of projects, and the team assesses the risks and determines if the project needs insurance in the form of risk response plans. The team also has a toolkit that has proven to enhance the probability of project success in the business context.

The project context is the environment in which the project takes place. The same project (same project core and risks) may be very successful in one environment and significantly challenged in another. The project manager, together with the project team, assesses the project context and addresses significant challenges in the environment.

The project context includes supports and enablers to the core project, including stakeholders, communications, and teaming.

1.9 Stakeholders, Communication, and Teaming

Projects involve people. Some of these people are customers, some sponsor the project providing money or resources, and the team does the planning and the project execution. All these people have a "stake" in the project; they are project *stake*holders. Good and effective communication is essential for these stakeholders to work well together and deliver successful projects. Finally, bringing people together does not insure that they work well together and deliver results. The project manager ensures that effective teaming up occurs, and stakeholders work well together.

1.10 Connect the Dots (Integration)

As a final step in project planning, the project manager ensures that all the pieces fit together as shown in the jigsaw puzzle. If the project manager finds that there are disconnects between pieces, he must adjust the plan so the puzzle pieces fit together.

Key stakeholders sign off on the project plan. When the team follows the process and keeps key stakeholders involved, sign off is a formality. If the project manager doesn't involve the stakeholders or doesn't get sign off before executing the plan, he risks alienating stakeholders, doing the wrong thing (project with features/functions that don't meet stakeholder needs), or doing the right thing, but doing it wrong (too costly, too slow, lack of quality).

The project manager obtains agreement and sign off on the overall approach before proceeding.

1.11 After the Plan: The Team Must Execute

Once the team has the project plan, it is ready to execute and implement. If there is a detailed plan, execution is simply a matter of following the steps listed in the plan.

While the team is executing, it is doing the work described in the plan. Executing takes longer than the time to create the plan. While executing the plan and doing the work necessary to create the product of the project, the project manager and project team continue to do project work that facilitates success, such as team meetings, communication, and teaming. Once the project is close to completion, the project manager begins to close down the project itself. The project manager and project team deliver the product of the project to its customers, they review the project experience in the form of lessons learned, and they move on to other projects and activities.

1.12 We Learn Best by Doing

Have you ever gone to a lecture or listened to a webinar and are unable to recall the key points of the session? Sometimes, we are unable to even recall the topic! Similarly with reading books—how often do you complete a book, or even a chapter in a book, and are unable to describe what you have just read. Listening and reading are passive approaches to learning; we sit back and let information pour into our heads. It's relatively easy, doesn't require much active engagement on our part, and as a result, it is not very "sticky." We are not able to recall much after a short time.

There is a Chinese proverb that states "Tell me and I'll forget; show me and I may remember; involve me and I'll understand."

As we get more engaged in the learning process, we tend to remember more of the content. This book offers alternatives with respect to the level of engagement with the topics and processes. The reader may simply read the book as a passive activity. The book includes a deck project as an example throughout the chapters and a case study, the Community Gardens (CG), with completed examples of the project plan and execution as appendices. The reader may simply read through these examples and apply them to his experience. We recommend that you actively engage in the topics, taking a critical look at the documents, standards, tactics, and processes. The book offers checkpoints at the end of sections where the book challenges the reader to pause and complete the work *specific to a project that the reader has selected.* If the reader takes this latter approach, he will have a completed project plan, ready for execution, when he finishes the book.

1.13 Checkpoint

Read the Community Gardens case study in Appendix A.

The reader should think about a project that he might use to practice the documents, standards, tactics, and processes introduced in the text. At the end of the next section, the reader begins work on the project of his choosing.

2
Project Charter

Stakeholders discuss many ideas for possible projects; not all these ideas become projects. Ideas become projects when the beneficiaries of the ideas commit to the planning and execution of the idea. This commitment takes the form of a formal or informal contract. Informal contracts are often incomplete, unclear, verbal, and subject to misunderstanding and inaccurate recollection. Formal contracts are written and approved in writing by the stakeholders doing the work, and the stakeholders paying for the work. A project charter is similar to a formal contract; it solidifies the relationship among key stakeholders and documents the work to be done at a high level.

A project charter documents the commitment to the project, defines the project at a summary level, and commits the resources necessary to deliver the project. Sponsor approval of the charter authorizes the project manager to proceed with the project; sponsor disapproval of the charter means that there will be no further work on the project.

Examples of projects include the following:

- Things
  - Complete an addition on a house
  - Purchase a car
  - Plant a garden (actually is more than planting—it's planting, purchasing, maintaining, and harvesting)
- Successful events
  - Wedding
  - Trip or vacation
  - Acceptance at a college of your choice
  - Investment in stocks or bonds
  - A community event, such as a craft fair or fund raising for local organizations
  - Evening out for dinner and the theater
  - Sale or purchase of a home

- Significant change in the environment or circumstance
    - Move into a new home or to a new country
    - Learn a new language
    - Change jobs

Not all of these projects have the same impact if successful. For those projects that cause a great deal of pain if they fail and for those projects that offer great opportunity if they succeed, it is worth the effort to develop a project plan and execute it. The key is whether the project objectives are significant enough to justify the effort; whether they are "plan-worthy." Generally, projects are "plan-worthy" if the pain of failure is greater than the effort of putting a plan in place.

2.1 Deck Project

Since it is challenging to discuss projects and project processes in the abstract, this book uses the addition of a deck to a home as a sample project. One scenario is a deck for the project manager and his family for a home he owns. If he and his family are flexible as to how big the deck is, how fancy the materials are, and are not concerned with completion date and cost, then he probably doesn't need a plan.

A second scenario might be a rental home, and the project manager needs to consider the owner's requirements. The owner may have specific requirements as to the size, the quality of the building materials, and may be concerned that it meets building codes. If the owner will reimburse the project manager for the cost of materials, he needs agreement on the maximum that he may spend. Deck completion date may not be important. In return for the effort and time spent constructing the deck, the owner agrees not to raise the monthly rent. However, the owner does plan to increase rent for the property based on the added value from the deck for future tenants.

A third scenario might be if a project manager agrees to build the deck for a neighbor. Similar to the landlord situation above, he needs agreement on scope, quality, and budget. In addition, the neighbor has family coming to stay from out of town and must have the deck completed prior to their visit. In this case, the completion date is also of concern.

In the first scenario (the project manager and family), there is little risk with respect to the scope, size, quality, budget, and completion date. In the second scenario (with the landlord), the landlord will not pay the project manager if the deck doesn't meet his expectations and even if it does meet his expectations, he will not pay more than the amount originally agreed upon. If the project manager is willing to assume the risk of nonpayment or nonreimbursement for all expenses, then he may not need a plan. If these risks are a concern, then the project manager minimizes the likelihood of this happening by developing a plan and obtaining the landlord's agreement to this plan in writing. The same situation applies to building the deck for a neighbor. In addition to concerns with the scope, quality, and budget, the project manager needs to be able to complete the work by the specified deadline. This requires planning and schedule calculations before the project manager agrees to take on the project.

The sponsor provides resources, in the form of money or people, to complete the work of the project. In the deck example cited above, the homeowner (or landlord) is the primary sponsor, as he pays for the deck and owns the finished product. The project manager ensures that the sponsor provides the following information:

- The project objectives
- The key people involved
- The date by which the project must be finished
- Any resource limitations in the form of people, money, or other resources
- Any quality standards that apply (e.g., building codes)
- Project success and acceptance criteria

This information should be as complete as possible and will be updated as the project progresses. As the project planning continues, more information becomes available and the project manager updates the initial information.

2.2 Project Charter Contents

The project sponsor and project manager use the project charter to determine if they should proceed. They need the following information to make this determination:

- Project purpose and requirements
- Objectives and success criteria
- Description and boundaries
- Key risks
- Assumptions and constraints
- Milestones
- Budget
- Stakeholders
- Acceptance criteria
- Approvals and date

These sections of the project charter include the following information:

Project Purpose and Requirements: The project purpose answers the question of why the stakeholders are doing this project. Sometimes, the purpose influences the design of the project. For example, if the primary purpose of the deck is to grow and display flowers or vegetables in planters, the project manager considers the amount of sun and shade. If the primary purpose is for reading and computer activities, the project manager might consider a canopy to ensure enough shade for these activities.

The requirements identify what the project and stakeholders need from the product of the project to meet the project objectives. The deck may need space for grill, a table, and chairs in order to fulfill the objective of spending more time outdoors.

Objectives and Success Criteria: In this section, the project sponsor and project customer identify what the project must accomplish to be successful. In some cases, they may not be able to determine project success until the project is complete, and the consumer of the product of the project has an opportunity to use the product. The objective is descriptive while the success criteria allow measurement of the extent to which the project has met its objectives. For example, the objective of the deck might be to spend more time with family outdoors; success criteria might be a minimum of 2 hours per day or 12 hours each week spent using the deck as a family. Often, the satisfaction of the success criteria cannot be assessed until the project is finished and the stakeholders are using the product of the project.

Description and Boundaries: Stakeholders expect that the project description supports the objectives and requirements.

The project manager identifies project boundaries by specifying what is out of scope. Listing out-of-scope items early in the process provides an opportunity for discussion as to whether selected items are part of the project or beyond the boundaries of the project. It is better to have these discussions early in the process so that all stakeholders have the same expectations. For example, the project charter (Figure 2.1) for the deck states that railings are out of scope. Since a raised deck without railings is likely to pose a safety hazard, this restriction implies a deck that is low to the ground. The charter also describes the deck as a three season deck; this suggests protection from excessive heat or cold.

Key Risks: Early identification of project threats and opportunities allows an early assessment of what might go wrong and provides opportunities to improve project outcomes. Project stakeholders use this information to make an early go or no-go decision for the project.

Assumptions and Constraints: An assumption is a statement that is considered to be true for planning purposes. Although the project manager doesn't know if the statement is true or not right now, he will find out over time. Assumptions either prove to be true, in which case, they are facts and the project manager removes them from the list of assumptions, or they prove to be false, in which case, they may be a risk to the project, and the project manager adds them to the list of risks.

Constraints limit the project team's options. If the deck must be at least 5 feet from abutting property, the team may not be able to situate the deck in certain places.

Milestones: This is the first attempt at a project schedule and tests the feasibility of delivering the product of the project within the required time frame. If the team cannot start on the deck

Project Charter

Project Name: Deck for Rental Property

Project Purpose and Requirements: Create an outdoor space to enjoy family activities during nice weather; enough room for grill, table, and 6 chairs; easily accommodate 10 people

Objectives and Success Criteria

- Objective—spend more free time outdoors
- Success Criteria—used on average 5 days per week
- Success Criteria—enhances rental value by 2% over equivalent properties

Description and Boundaries

- 3-Season deck, wood floor, aesthetic fit with house
- Out of scope—railings, lighting, stairs

Key Risks

- Threat—soil cannot support weight of deck; need to "reinforce" the soil at extra cost/time
- Threat—neighbors concerned about loud parties
- Opportunity—abutting neighbors hear about the idea and want to build decks also; opportunity for reduced costs in case of bulk purchases

Assumptions and Constraints

- Assumption—owner secures any necessary building permits prior to construction
- Constraint—sides of deck must be at least 5 feet from abutting property

Milestones

- Kickoff and charter approved—Project start
- Permits secured—week 1
- Materials purchased—week 3
- Construction begins—week 4
- Construction complete—week 5
- Town inspection complete; approved for use—week 6
- Lessons learned and project completion—week 6

Budget

- Costs: building supplies, permits—$3,000
- Revenue—none
- Note that grill and deck furniture are out of scope for this project

Stakeholders

- Project manager, property owner, renters, neighbors

Acceptance Criteria

- Property owner signs off on project; decides if project meets the requirements
- Renters' evidence approval via use of the deck

Approvals and Date

Project Manager: ______________________ **Date:**____________________

Sponsor (Property Owner): ______________ **Date:**____________________

Figure 2.1 Project charter for deck project.

until August or September due to other commitments, it may make more sense to delay the project until the following spring. Milestones are significant events; they do not address the process to get work done, they identify the dates by which work must be complete.

Budget: This is the first estimate of costs and revenue, if any.

Stakeholders: This is the list of people involved with, or impacted by, the project. It is a best practice to keep the most influential stakeholders involved in project planning and execution.

Acceptance Criteria: This is slightly different from the success criteria listed earlier. The success criteria focus on whether the project satisfies the project purpose and may require evaluation many weeks or months after the project completes. Acceptance criteria focus on conditions *at the time* the project completes; it may be less stringent than the success criteria. The project sponsor and customer may be willing to accept a project that is less than 100% successful, such as a deck that is a bit smaller, more expensive, or completed late. The acceptance criteria are the conditions that, if met, then appropriate stakeholders sign off on the project, make final payments, and accept ownership and responsibility for the ongoing maintenance of the product. The project customer defines project acceptance criteria; the project manager documents this project acceptance criteria.

Approvals and Date: Both the project manager and sponsor must sign the project charter indicating their support and approval of the effort.

Figure 2.1 illustrates a project charter for the scenario of building the deck for the owner of rental property.

Appendix B contains the project charter for the Community Gardens project.

2.2.1 Checkpoint

At this point, you need to begin your project plan. Select a project that you consider "plan-worthy." Use the sample charters for the deck and the Community Gardens projects to guide your efforts. Write down the preliminary information required for the project charter. Obtain the appropriate sign offs on this information.

Once the project charter is complete and approved, the project manager further identifies the work of the project in the form of the project scope statement.

3
Project Scope

Once the project manager has an approved project charter, he is ready to develop the details of the project core. As illustrated earlier, the project core is like a jigsaw puzzle with four pieces (scope, schedule, budget, and quality), and project risk bounds this core project. Project scope and project quality determine how long a project takes and how much a project costs; project risk influences the ease or difficulty of delivering the scope and quality on schedule and within budget.

The journey through project core begins with an exploration and elaboration of project scope from the preliminary information in the project charter.

Think of the project charter as a preliminary assessment before committing to a larger project. For a medical issue, it is common to have x-rays or MRIs to determine if more extensive medical treatment is warranted. People often request an estimate of repairs for a car or appliance before committing to larger and more expensive repair work. In business, companies often build a prototype before committing to full scale development and production. And, homeowners often request a proposal from contractors before committing to extensive renovations.

In the project scope, the project manager expands on the information in the project charter and answers the questions of "who, what, why, where, when, and how." He validates the preliminary go or no-go decision made based on the project charter and aligns key stakeholders around the project core. Once he has a sign off on the project scope statement, he creates a work breakdown structure (WBS) to capture and organize the key project and product deliverables.

3.1 Scope: Tele*scope* to Micro*scope*

The project manager starts his definition of the work (the "what") with a big picture view and then he "drills down" to develop a more detailed view of the work. He first looks at the work of the project through a telescope—it is far distant with few details, but he has a general idea of the results of the project. Over time, he increases the magnification to see more details. At the end of the planning process, he is effectively looking at the work through a microscope, so detailed that he looks separately at the individual pieces of work. Going from tele*scope* to micro*scope*, the project manager calls this scope—the project scope. Everything that needs to get done is "in scope"—he needs to be able to view it through the telescope initially and later examine details through the microscope. Anything that is excluded from the project is "out of scope." These things are not visible through the telescope. And he does not look further at anything not visible through the telescope.

When the project manager uses the telescope, he aims it far away and takes in the big picture. He's not concerned yet with details of the "what," and he does not focus on how he might accomplish the work.

A pilot looking to land an airplane is certainly looking at the airport. He's also concerned with what surrounds the airport. At the end of the runway, is there a highway, a field, or a harbor? He's also aware of the cloud cover, the wind (as evidenced by wind sock) and the wind speed (as evidenced by the anemometer), any hazards (other planes, trucks, or fire equipment) on the runway, and any significant hazards in the taxi areas.

The pilot needs to take this big picture and break it into smaller, more discrete pieces. In the example above, the big picture is "Land the Plane"; the pilot breaks this big picture into smaller parts in order to determine the "how." Some logical categories might be the plane, the runway, the airport, and the weather. There is no absolute right or wrong list of categories; the categories set the stage for determining the details of what else is necessary for a successful landing.

Stores are organized by categories of goods. In department stores, there are sections for shoes, women, children, men, and housewares. In grocery stores, merchandise is organized by meat, vegetables, bread, crackers, etc. And there are some areas like frozen foods, where products cross all these categories. Hardware stores organize by rooms (kitchen, bath, etc.) and then some categories that cross these specific room categories like paint and lighting.

Why chose this type of organization? What if stores organized their products alphabetically? Imagine how difficult it would be to find things. This is also impractical—grocery stores need to keep all frozen foods together and all foods that require refrigeration in the same place. It would be reasonable to maintain organization by freezer or refrigerator for logistical reasons but organize alphabetically within this larger structure. The store might have tuna fish next to taco shells and tomatoes, and crackers next to crabmeat. Or, the store might organize alphabetically by brand names, Heinz ketchup, pickles, and mustard are co-located, and the same products by another vendor are in another section with all products by that vendor. Unfortunately, these approaches to organizing the contents of a grocery store do not make it easy for customers to find the items they wish to purchase.

Project managers have a similar challenge with their projects. If the telescopic view details everything, it is difficult for the stakeholders' minds to make sense of the big picture. It seems to be a random set of discrete items and too many individual pieces for their minds to track. So he needs to find some reasonable organization to break his telescopic view into smaller, reasonably logical microscopic views. This organization facilitates the decomposition of the big picture into smaller, more manageable pieces.

3.2 Project Scope Statement

The project scope statement elaborates on the information in the project charter. During the creation and approval of the project charter, the project manager talks to the key stakeholders to understand their needs, expectations, and details of both the project and the product. The project charter addresses the entire project; it examines all components of project core (scope, schedule, budget, and quality), risk, stakeholders, and other significant considerations.

The project scope statement focuses only on "the what" of the project. The project manager drills down for details of the scope; he considers the other project aspects (schedule, budget, quality, and risk) once scope is defined and approved.

At a minimum, the scope statement includes the following information:

- Scope description
- Acceptance criteria
- Deliverables

- Exclusions
- Constraints
- Assumptions
- Approvals and date

While the project manager captures much of this information in the project charter, the level of detail in the charter is sufficient only for a go or no-go decision for the project. The project stakeholders need more information to develop a detailed plan for the project.

Scope Description: This section provides a description of the product, in this case, the deck. The focus is on "the what," not how to accomplish the work. More details clarify stakeholder expectations and increase the likelihood of a satisfactory product. For example, if the color of the wood for the deck is important, the project manager must specify it. Otherwise, he may get blond wood or a dark mahogany. An effective approach is for the project manager to document the requirements as if someone else will do the work; he lists the details necessary for this other person to meet his expectations.

Acceptance Criteria: This is a further elaboration of the acceptance criteria listed in the project charter. The project manager talks to stakeholders to completely understand the project and product scope; as he gathers this additional information, he is better able to articulate the minimum criteria for the customer to approve and accept the product of the project.

Deliverables: Deliverables are the "what" of the product and the project. This is a list of all the things (products or services) that project stakeholders have once the project completes. The project manager further characterizes project deliverables in the Work Breakdown Structure (WBS), discussed below. He creates a detailed list of the deliverables, collects and organizes them as a graphic to ensure they are complete, and documents them in an outline form for sign off and distribution.

Exclusions: This section builds on the out of scope segment of the "Description and Boundaries" section of the project charter. While the project charter addresses exclusions to all aspects of the project (e.g., budget and quality), this section of the project scope statement highlights exclusions specific to project and product scope.

Constraints: A constraint is a factor that restricts project execution. In the project scope statement, the focus is on constraints specific to project and product scope. An example might be a requirement that only environmentally friendly raw materials may be used.

Assumptions: These are items assumed to be true for planning purposes. As noted earlier, these will either prove to be true (so no longer an assumption) or prove to be false and be moved to the list of risks.

Approvals and Date: The project manager ensures that the project scope statement is signed and dated by key stakeholders, including at a minimum, the project sponsor and customer. Although the project manager likely authored the project scope statement, he should sign and date the document indicating his approval and acceptance.

Figure 3.1 illustrates a project scope statement for the deck project.

Appendix C contains the project scope statement for the Community Gardens project.

3.2.1 Checkpoint

Using the sample scope statements for the deck and the Community Gardens projects, prepare a project scope statement for your project. Work with your team members and share the draft scope statement with key stakeholders. Once you complete your project scope statement, include it in your project plan. The project scope statement is the basis for the WBS, discussed below.

3.3 Work Breakdown Structure

The WBS is the basis for other project components such as schedule and budget. The project manager estimates the schedule and budget for all deliverables in the WBS; if a deliverable is not in the WBS, he does not likely have the schedule and budget estimates for that work.

Project Scope Statement

Project Name: Deck for Rental Property

Scope Description

- Minimum dimensions 12 feet by 15 feet
- Wood planks with appropriate drainage
- Aesthetic fit with house

Acceptance Criteria

- Town inspection certificate received
- Property owner signs off

Deliverables

- Deck
- Approved permits and inspections
- Project management documentation (e.g., charter, scope statement, etc.)

Exclusions

- Railings, lighting, stairs
- Deck furniture, grill, flowers, or other plantings

Constraints

- Minimum 5 feet away from all abutting property
- Compliance with building regulations
- Complete and usable by mid-summer (so can enjoy during current lease)

Assumptions

- Property owner secures any necessary building permits and inspections
- Abutting neighbors will not object

Approvals and Date

Property Owner as Sponsor: ____________ **Date:** ________________

Project Manager: ____________________ **Date:** ________________

Figure 3.1 Project scope statement for deck project.

As a result, the project will be delivered late or over budget or he eliminates some of the scope to deliver on time and budget. If he finds that he has missed a deliverable during the planning process (and before the project plan is approved), he modifies the WBS and includes the necessary schedule and budget. If he discovers that a deliverable is missing from the WBS after the project plan has been approved, he initiates a change to the WBS to include the deliverable, along with the necessary schedule and budget.

The project manager frames project scope such that he is able to estimate schedule, budget, and quality. Is "building a deck" a sufficient level of detail to estimate schedule, budget, and quality? He likely needs more detail, such as the size of the deck, to accurately estimate this project.

There are many ways to organize the scope of a project. Best practice has shown the most effective organizing principles are by deliverables or by chronology as determined by project phase. Deliverables are the things (products, services, results) that the project manager plans to have when he completes the project. In the example of the deck above, the deliverables would be a deck and any documentation developed in support of building the deck. The deck is an example of a product deliverable; the documentation is project deliverable.

Project phases are logical groupings of activities. For example, a generic project life cycle contains four phases:

- Starting the project
- Planning the project
- Executing the work of the project
- Closing the project

The most effective project organizing principles are those that enable the project team to completely identify the work of the project. The work of the project is the "what"—what the stakeholders will have when the project is complete. It is *not* the "how"; the process for accomplishing the work of the project. The project manager completely identifies the work of the project and obtains agreement on this work *before* he spends time and resources determining how to accomplish this. Suppose the project manager determined the schedule and budget for building the deck *before* he obtained agreement on the scope. If he assumed that the deck would have a railing, he would spend time estimating how long the railing would take and how much it would cost. If he later found out that the deck is only one foot off the ground, and requires no railing, the time spent analyzing the schedule and budget for the railing is wasted effort. Getting approval on the "what" before he spends time on the "how" avoids wasting time on things that are "out of scope."

3.4 Decomposition of the Deck Project into WBS Categories

The major deliverable of this project is a deck completed according to the specifications in the scope statement. However, there are other deliverables. For example, the project needs to complete the appropriate safety and regulatory deliverables. The project manager needs to complete the necessary project management documentation and services. If he only considers the building of the deck, he may overlook some of the other items. If he fails to secure the necessary safety and regulatory permissions, he may not be able to use the deck until the town grants the permits. This might simply delay his use of the deck or it may require changes to the deck that will be more costly to implement after the deck is complete than it would have been if he addressed the issues while constructing the deck. If he fails to complete the project management documents, he may build a deck that is too small, doesn't meet the property owner's requirements, or he might underestimate the schedule and budget to complete the deck.

Figure 3.2 identifies four deliverables: materials and supplies purchased, deck constructed, safety permits and inspections complete, and project management documentation and services complete. Once he gets approval for these deliverables, he identifies the tasks necessary to complete these deliverables. This WBS is the "what" of the project; the "how" comes later.

The project manager may also display the WBS as an indented hierarchy, often called an outline, as follows:

Deck WBS as indented hierarchy or outline

1.0 Materials and supplies purchased
2.0 Deck constructed
3.0 Safety permits and inspections complete
4.0 Project management documents and services complete

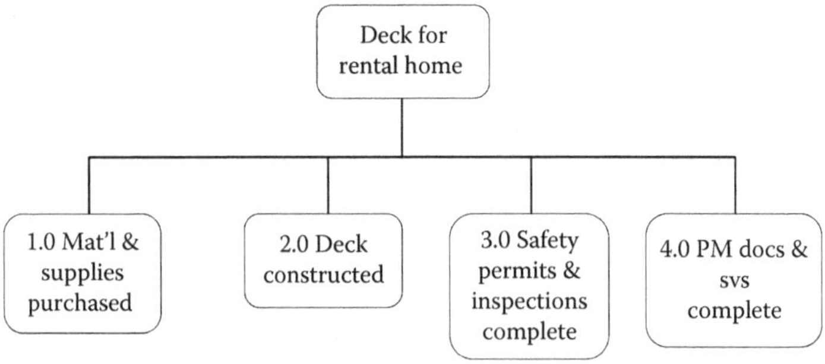

Figure 3.2 WBS for deck project—Graphical format.

There are relatively few deliverables for the deck, and the WBS is straightforward. Many projects are more complex, with many deliverables and alternatives as to how to best structure the WBS. The WBS for the Community Gardens project in Appendix D illustrates a WBS with more than one level of detail. The lowest level deliverables in a WBS are "work packages." As these work packages are deliverables, it is a best practice to phrase them as if they are already complete. For example, the WBS states "Deck Constructed" rather than "Construct the Deck." Deck constructed connotes a completed deliverable; that is, the deck is complete and ready for use. Construct the deck implies a process of building the deck and is more appropriate to that section of the project plan where the project manager discusses the "how" of the project. By using the "noun–verb" phrasing in the WBS, he is defining a future state—the "what" or the deliverables of the project.

The top levels of the WBS are numbered consecutively, by deliverable or chronology based on the organizing structure of the WBS, in the format of 1.0. If there are multiple top levels, as in the Community Gardens project, the top level 4.0, pre- and post-season preparation, is further decomposed to 4.1, pre-season preparation and 4.2, post-season preparation (Figure D.2). Tasks within a WBS category continue the numbering scheme. For the deck project, WBS category 1.0, materials and supplies purchased, is further decomposed into tasks 1.1, 1.2, and 1.3 as illustrated in Figure 4.2.

The planning effort should be appropriate to the risk entailed if the project manager fails to deliver the product of the project. This planning is a balance of effort and risk. The risk involved in building a deck for a rental home is relatively small. If the team neglects to purchase some of the necessary materials, generally they return to the store and obtain whatever is necessary. But, what if the rental home is in a remote location many miles from the nearest stores or so remote that access is limited to small planes or boats? In these situations, the risk of neglecting to secure all necessary materials is far more significant. Another example is a hiking trip. The impact of forgetting some supplies for an afternoon hike is far less significant than forgetting supplies for an expedition to climb Mt Everest. Failing to bring appropriate food, drink, oxygen, and camping equipment can mean death or serious injury to the participants in the expedition. The effort of planning is quite reasonable given the consequences of not planning!

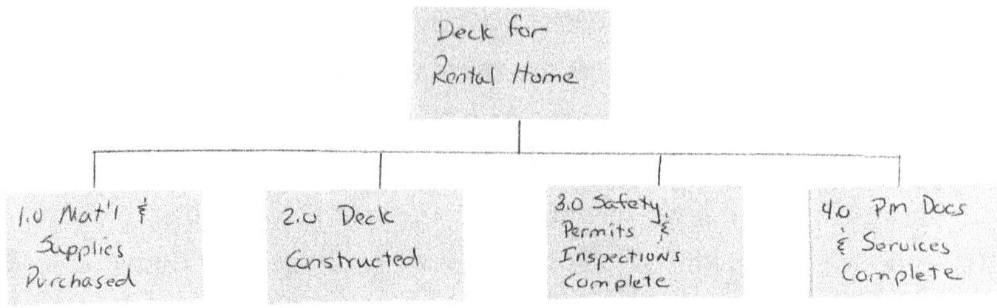

Figure 3.3 WBS for deck project—Sticky note format.

Creating the WBS helps the project manager and the project stakeholders refine the telescopic view of the project. The process of discussing the project and identifying the major components force them to think about the best approach. The components of the WBS and the groupings are likely to change as the project manager goes through the process of building and getting agreement on the WBS. Using sticky notes to capture the WBS components is an excellent tactic for engaging other team members in the process of defining the project and facilitating changes while defining the project and its components. Figure 3.3 illustrates the WBS for the deck project using sticky notes.

Appendix D contains an example of the WBS for the Community Gardens project. It includes the WBS as a graphic using sticky notes, as a typed (graphic) tree structure, and as an indented outline.

3.4.1 Checkpoint

Now, it is your turn. Refer to your project plan and refresh your thinking about your selected project. Think about the telescopic view and set this up as your top level WBS. Then think about the component parts—the "what" or the work packages and write these down. Note that your first pass at this work may not be the final list of work packages. The initial picture provides the basis for expanding the level of detail. As you and your team move into some of the additional details, you may find that you need to expand or collapse the list of work packages. This is fine and normal. These pieces are a "work in progress" until you finish planning and obtain sign off on the total effort. Use this as an opportunity to generate consensus and comments from the other people involved in the project.

Using your project scope statement, build a WBS using sticky notes. Work with your team and share it with other key stakeholders. Once it is complete, organize it into an indented hierarchy and include these documents in your project plan. You may also choose to type up the WBS as a tree structure graphic and include it in your project plan.

3.5 Next Steps

Once the project manager has an agreement and sign off on the WBS work packages, he moves on to the tasks that comprise the work packages. He "extends" the work packages by adding the tasks necessary to complete the work packages. This is the "how" of the project and enables the development of a project schedule. Once the project manager knows how long the work will take, he can estimate a completion date. The next chapter addresses the work of completing the project deliverables.

4

Project Schedule

Now that the stakeholders have defined the project scope, the project manager focuses on the timing and schedule for the project. Working with the team, the project manager expands the work packages (the what) and identifies the "how" for each of the detailed work packages. They list the tasks necessary to complete the work package deliverable and sequence the tasks. Some things need to happen before others (for example, the team needs to put in the foundation footings for the deck before they can put down the surface planks, and they need the surface of the deck completed before they can put up railings, if railings are in scope). Some things can happen at the same time (such as determining permit requirements and obtaining approval of schematics for deck).

For each task, the project team determines how long it takes, called duration, and how much effort it takes. The project manager also determines the resources—who is going to do the work, what unique skills are needed, how many people work on the task, and how much time is available for the work. In other words, are they available full time or part time.

The project manager assigns each work package to the person who will do the work. This team member identifies the tasks necessary to complete the work package. It is risky to have one team member identify and estimate the tasks when another team member will do the work. If one person identifies the tasks and task effort and another person will complete the work, the person doing the work may have another approach and anticipated effort. This makes it challenging to hold the people doing the work accountable to the planned schedule.

For the deck project, the team identifies the tasks necessary to having all the materials and supplies in hand so they can begin to build the deck. This entails creating a list of everything needed including details such as size, quantity, brand, and type. In the case of the wood for the deck, the team would specify dimensions of planks, number of planks, whether the wood is weather treated, color, and any other relevant information. The team would next shop for the best price and arrange for pick up or delivery of the materials.

The extended WBS for the deck project also lists tasks for the other deliverables, including deck constructed, safety permits and inspections complete, and PM documentation and services complete. If any of the tasks are not clear from the short description on the extended WBS, the project manager separately creates these descriptions and includes them in the project plan.

Best practice is to list all tasks, including task name and number, on sticky notes (similar to the process for work packages in the creation of the WBS) and put these sticky notes below the work package to which they belong. The sticky notes allow the team to easily add, delete, and change tasks until the team agrees on the completed extended WBS. It is easy to differentiate between work packages and tasks if the team uses sticky notes of different colors and/or size for the WBS work packages and the project tasks.

The project manager photographs the finished extended WBS and later may document it (in graphical format) using an appropriate computer application. He may also list the tasks documented in outline form under the appropriate work packages in the WBS.

Figure 4.1 illustrates the extended WBS for the deck project in sticky note format.

Figure 4.2 illustrates the graphical extended WBS for the deck project.

Table 4.1 illustrates the extended WBS for the deck project in outline format

Appendix E contains the extended WBS for the Community Gardens project in sticky note, graphical, and outline formats.

4.1 Checkpoint

Using the examples for the WBS extended for the deck project and the WBS extended with tasks for the Community Gardens project in Appendix E, prepare an extended WBS for your project. Work with your team members and use sticky notes to involve everyone in the process.

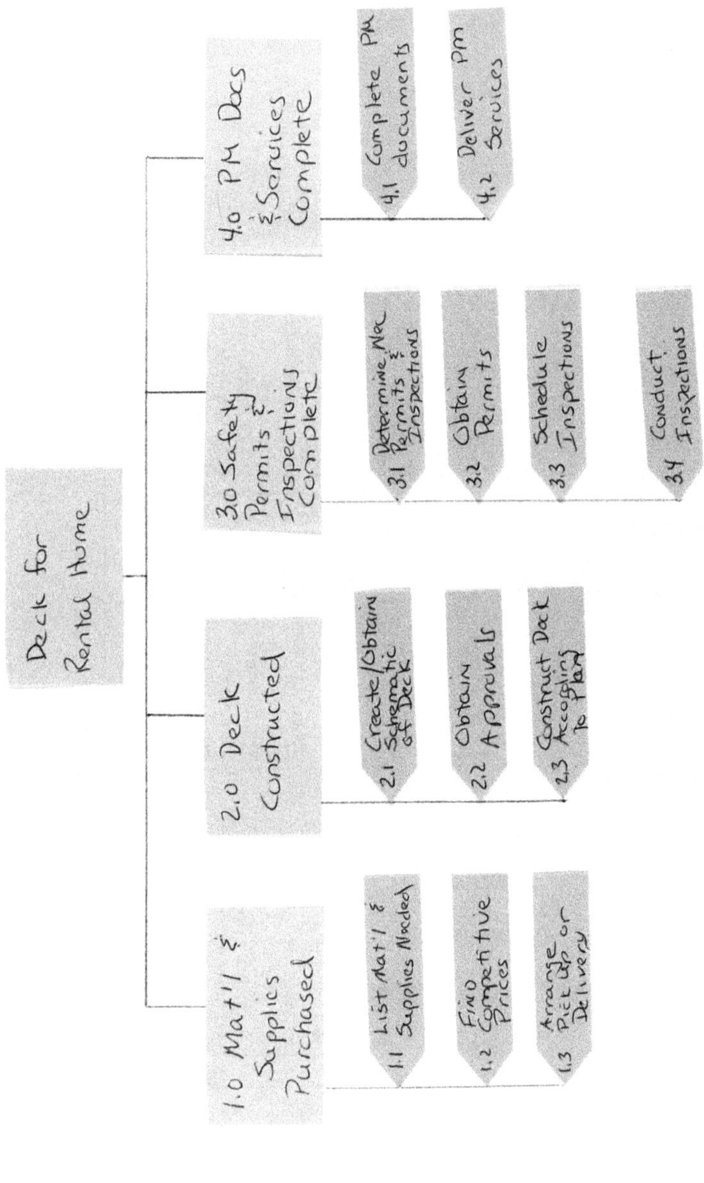

Figure 4.1 Extended WBS for the deck project—Sticky note format.

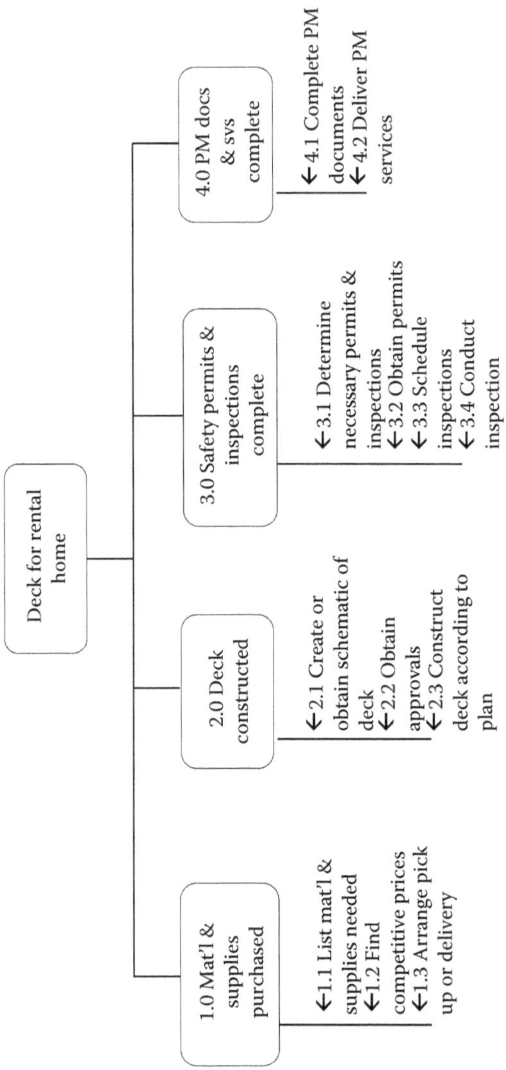

Figure 4.2 Extended WBS for the deck project—Graphical format.

Table 4.1 Extended WBS for Deck Project—Outline Format

1. Materials and supplies purchased
   - 1.1 List material and supplies needed
   - 1.2 Find competitive prices
   - 1.3 Arrange pick-up and delivery
2. Deck constructed
   - 2.1 Create or obtain schematic of deck
   - 2.2 Obtain approvals
   - 2.3 Construct deck according to plan
3. Safety permits and inspections complete
   - 3.1 Determine necessary permits and inspections
   - 3.2 Obtain permits
   - 3.3 Schedule inspections
   - 3.4 Conduct inspection
4. PM documents and services complete
   - 4.1 Complete PM documents
   - 4.2 Deliver PM services

Projects should use processes that involve the stakeholders so that they feel ownership for the project, project processes, and project deliverables. Stakeholders are more vested in the project when they view it as "our" project rather than "your" project. When they feel ownership, they will work harder to ensure project success. Getting stakeholders involved in identifying and later sequencing the tasks on sticky notes promotes shared ownership of the project.

Once stakeholders review and approve the extended WBS, you may type the information into a computer program for ease of sharing and distributing the information. Once the stakeholders approve the completed extended WBS, include it in your project plan. The tasks in the extended WBS are the basis for the schedule network diagram, discussed below.

4.2 Schedule Network Diagram

Once the team determines the work to be done (tasks) as extensions to the WBS diagram, they sequence these activities. They determine what should come first, what comes later, and what tasks can be done at the same time. For example, for the deck project, the team must acquire materials before starting to build the deck. The deck must be completed before inspections and acquisition of usage permits.

After identifying the tasks, the sequencing, the effort, and the resources, the project manager calculates how long the total project will take. He determines the earliest completion date (assuming nothing goes wrong) if the project begins on January 1. Or, he determines when the team must start if the project must be complete by December 31. If the finish dates don't meet the requirements as documented in the scope statement, then key stakeholders must agree to a later completion date, de-scope (eliminate some of the deliverables), or think about applying extra resources (so there will be more cost) to complete the project on original schedule.

This structured approach makes it difficult for the team to pretend that they can work faster/smarter/luckier and complete the project in a shorter, unrealistic amount of time.

The project manager follows a process similar to that for creating the WBS. He convenes a meeting of the team members responsible for the tasks and lists each task on a sticky note. The task sequencing is independent of the WBS categories; the tasks from different WBS groupings are intermixed. The task numbering ensures that the task association with the WBS groupings is maintained.

Figure 4.3 illustrates the schedule network diagram for the deck project in sticky note format.

Figure 4.4 illustrates the schedule network diagram for the deck project in graphical format.

While some of the tasks have mandatory dependencies, the relationship among other tasks may be discretionary. For example, the team must acquire the materials and supplies before they start construction of the deck; it is simply not possible to construct the deck without the supplies. This is an example of a mandatory dependency. The sequential relationship of task 3.2, obtain permits, and task 1.1, list material and supplies have some latitude. The project manager might choose to do these tasks in parallel (at the same time) rather than sequentially. Based on the sequencing in the network schedule diagram (Figure 4.3 or Figure 4.4), it is apparent that he chose to delay the listing of material and supplies until the permits are secured. If the team starts working the list of materials and supplies, and the permits are denied, then the time spent on the list is wasted and may have to be repeated based on different requirements. Permits might be denied if the size of the deck puts it too close to the property line

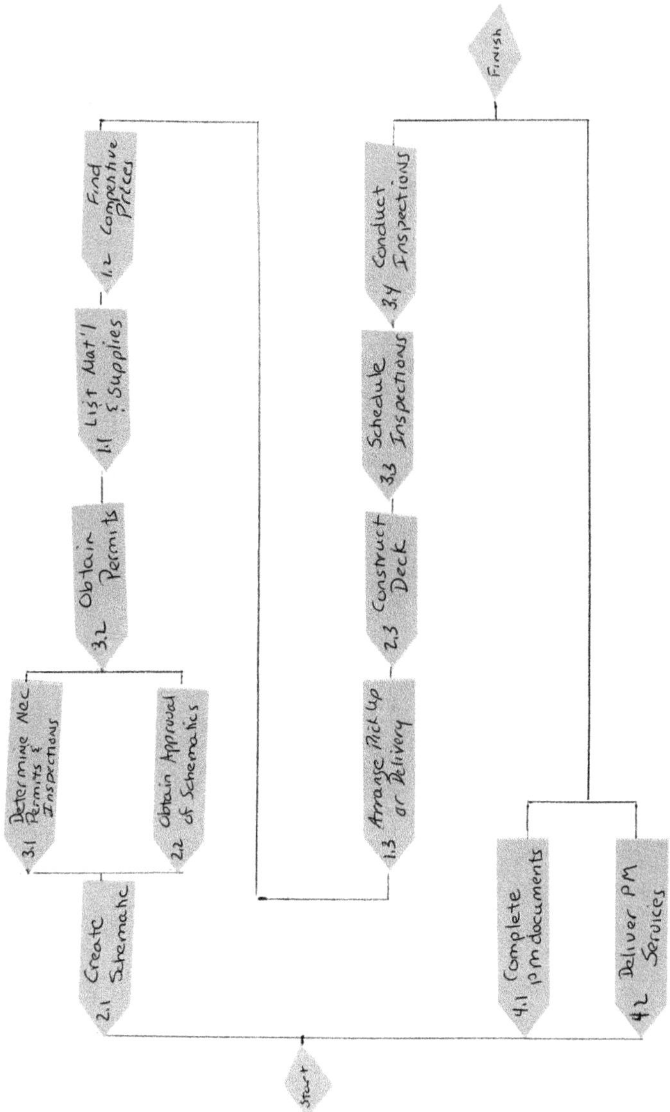

Figure 4.3 Schedule network diagram for the deck project—Sticky note format.

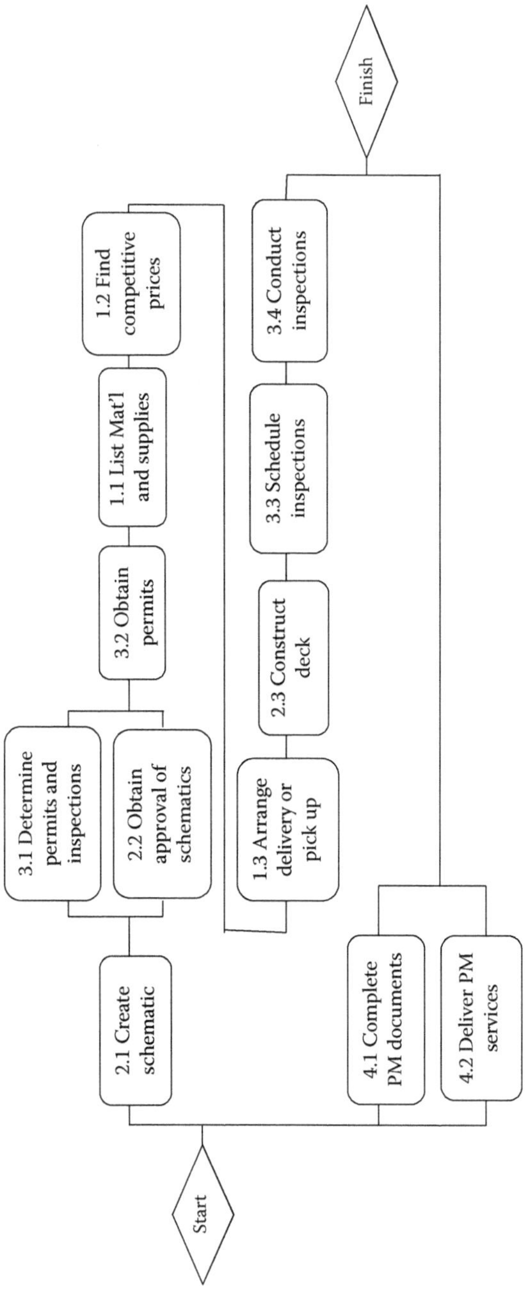

Figure 4.4 Schedule network diagram for the deck project—Graphical format.

to meet local codes. In this case, the project manager may need to propose a smaller deck to the sponsor; the sponsor may not approve a smaller deck. If the deck will be smaller, the list of materials will be different to accommodate this reduction in deck scope.

The project manager assesses the completeness of the schedule network diagram and the complexity of the project by identifying the paths from start to finish. For the deck project, these paths are as follows:

Start → 2.1 → 3.1 → 3.2 → 1.1 → 1.2 → 1.3 → 2.3 → 3.3 → 3.4 → Finish
Start → 2.1 → 2.2 → 3.2 → 1.1 → 1.2 → 1.3 → 2.3 → 3.3 → 3.4 → Finish
Start → 4.1 → Finish
Start → 4.2 → Finish

The project manager verifies that each task has a predecessor and successor, and notes that the first two paths are the most complex, and therefore have more risk. He pays attention to those paths with the most risk as they have the potential to delay project completion.

4.2.1 Checkpoint

Use the schedule network diagram for the deck project in the text above and the Community Gardens project schedule network diagram in Appendix F as samples for developing the schedule network diagram for your project. Include each task listed in the extended WBS on your schedule network diagram.

Write each task on sticky note and arrange the sticky notes in the order in which they will be completed. Remember that some items can be completed at the same time (in parallel) while others are sequential. Using sticky notes makes it easy to move the tasks around until you get a schedule network diagram that works best.

4.3 Project Duration

The project takes as long as the longest path through the schedule network diagram. To determine how long it will take to complete the project, the project manager identifies the paths through the diagram.

Once the team determines task durations, the project manager calculates how long each path will take. Below is an example of a simple project and the process the project manager uses to determine paths and durations.

In this example, three college friends plan a reunion in San Francisco. They have all relocated to different cities; although they could meet in one of the cities in which they live, they all really like Fisherman's Wharf in San Francisco and decide this would be an excellent place for the reunion. Peter now lives in New York. He likes road trips so he will be driving across the country. In order to be able to drive straight through, he'll bring a couple of friends to help with the driving. He calculates his trip at just over 3000 miles, and driving at about the speed limit, he calculates it will take about 42 hours to get to San Francisco. Mary currently lives in Topeka, Kansas; she has to travel about 1800 miles and expects the trip to take around 1 day (25 hours). She has back problems so needs to make frequent stops to stretch. Also, she plans to take Patch, her dog, so she will need to stop frequently for him. She will stay overnight for 1 or 2 nights for the trip. Paul lives in Phoenix, a mere 850 miles (12 hours) from San Francisco.

If everyone agrees to leave at 9 a.m. on Monday, Peter will arrive around 3 a.m. Wednesday morning (driving straight through, since there are multiple drivers in his car). Mary decides to limit her driving to 8 hours a day and spend 2 nights in hotels en route; she will arrive around 5 p.m. on Wednesday. Paul drives straight through and arrives at Monday at 9 p.m. Since everyone must be at Fisherman's Wharf before the reunion can begin, the earliest that the reunion can start is 5 p.m. on Wednesday when Mary arrives (Figure 4.5).

Peter will arrive 14 hours before the reunion at 5 p.m. on Wednesday. He can use this time to rest from the long trip, or he might decide to stay spend a night en route in a hotel to rest. Paul will have almost 2 days of free time. He might choose to do some sightseeing en route and arrive a bit later. This "extra" time that Peter and Paul have is called float or slack; it means free time available to some of the paths.

Since they have extra time, one of Peter's passengers volunteers that he has a friend who lives en route, and she is willing to put them up for the evening. As with any poorly scoped project, Peter and team decide this is a great idea and plan to stop en route, delaying their arrival by at least 12 hours and more likely by 18 hours since

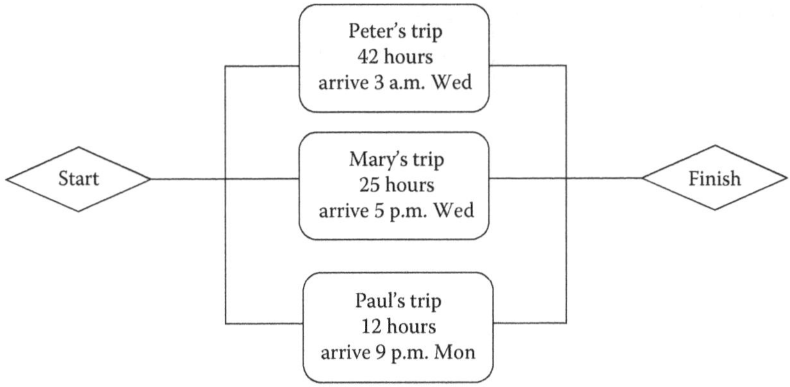

Figure 4.5 Project duration.

it is not polite to just stop by and sleep, but they need to spend time entertaining their host. So now he arrives around 9 p.m. Wednesday. Everyone agrees that they will push the reunion start to first thing Thursday morning.

Since this means that Paul will now be in San Francisco without his friends all day Tuesday and Wednesday, he decides to delay his departure until Wednesday at 9 a.m. Peter and Mary now start rethinking their departure times!

In the first scenario, the stakeholders decide on a project start date and calculate the earliest that they can finish the project. In the second scenario, they decide on a project completion date and calculate the latest that they can start all the work of the project and still finish on time. Either approach is fine.

The project manager applies this thinking to his projects. If he starts on March 1st, what is the soonest that the project will finish. If the project needs to be completed by September 15th, what is the latest he can start the project and still meet this date constraint. Here are some examples to consider:

- There is a wedding (or other key event) scheduled for June 16th. What is the latest the planning can start and still have everything done in time?
- Mortgage application is due on October 2nd (or the home buyer loses the lock in rate). What is the latest the prospective homeowner can begin the application and collection of backup data and still complete the application on time?

- If the project manager starts building his deck on June 1st, when will it be complete so he can schedule his first cookout on the deck?

The project manager then validates this schedule against the preliminary estimate documented in the project charter and decides if this project still makes sense.

4.4 How Much versus How Long

There is another consideration before the project manager finalizes the project schedule. He needs to consider "how much" versus "how long." This differentiates between the concepts of schedule (how long) and budget (how much). Consider the situation of building a deck and assume the project manager hired a contractor to do some work. He tells the project manager that it will take 40 hours of effort to do the work. Assuming 8-hour days, this will take one 5-day week. But, what if the contractor is already working on another job and he can only work on the job half time. This same 40 hours will now take two 5-day weeks (assuming 4 hours each day). If he is really busy with 4 other jobs, he can only work 20% of the time on the project so the project manager only gets one 8-hour day each week. In all cases, the project manager pays the contractor for 40 hours (assuming that is the time worked). But, depending on the percent of time that the contractor works on the project, the duration of the project (and by extension the completion date) will extend further into the future. So the effort determines the cost; the percent of time available determines the duration and schedule. Project managers sometimes call effort "applied time" and call duration "elapsed time." Strictly, elapsed time includes weekends and other nonwork time (such as holidays). As in the case above of the contractor working 2 jobs simultaneously and only available 4 hours per day for the job, the elapsed time is 10 days. If the contractor takes weekends off, the real elapsed time will be 12 days.

In some situations, project duration is a function of repeating tasks, rather than the sum of durations and time spent waiting. In the Community Gardens case study, project completion is a consequence of weekly harvests. The project manager doesn't need to calculate wait times between weekly harvests; he simply needs to know when the team will stop harvesting and then add the "end of season" tasks (such as 4H competition, year-end cookout, and lessons learned) to the last harvest date to determine project completion date.

4.5 What about Resources?

In order to determine how much effort (applied time) a task requires, the project manager considers the expertise of the person doing the work. To determine how long the task will take (elapsed time), he considers the number of people working on the task. If the project manager plans to build the deck himself, and this is the first deck he has ever built, it will take him longer than someone who is an expert at building decks and builds decks as a full time job. If he plans to build the deck by himself, and he plans to work on it full time, it will likely take 40 hours, or five 8-hour days to complete. If his neighbor helps and he also is willing to work full time, then the two of them will complete the job in 40 hours, but over 2 1/2 8-hour days. If he hires an expert contractor, he might be able to complete the deck in 32 hours or four 8-hour days.

Before the project manager calculates the schedule, he augments the schedule network diagram with information about work hours (effort), percent time, nonworking time, and resources. The work hours differ depending on the expertise of the resources. The project manager considers the resource expertise when developing the estimate for work hours.

4.6 Building the Project Schedule

To create the project schedule, the project manager builds on the sequence of tasks in the schedule network diagram. For each task listed in the schedule network diagram, the project schedule includes applied time, percent of time, duration, and resources (both number of resources and skills/specialties if appropriate). The project manager may also track both working time and calendar time. Working time is the time spent on project activities; calendar time includes weekends, holidays, and other nonworking time. Working time identifies the number of days required to complete the project *if there was no nonworking time.* Calendar time adds weekends and other nonworking time and extends the completion date of the project. If project resources worked every day, working time and calendar time would be the same. In the real world, it is common to take weekends and holidays off, so the number of working days is generally less than the number of calendar days and extends the projected project completion date. The project schedule includes rows for counting both working days and calendar days. The deck project takes 40 calendar days but only 30 working days.

Looking at Figure 4.6, there is the task list along the left side of the chart. For each task, the project manager includes values for applied time (effort), percent time, duration, and resource. He schedules the tasks in the columns for work days and for calendar days. Task 2.1 (Create Schematic) takes 4 hours; since it begins on day 1, the project manager puts an "x" in that cell. Since it only takes 4 hours and the resource is working full time, he might choose to start tasks 3.1 and/or 2.2 (these tasks follow task 2.1 as noted in Figure 4.4) during the afternoon of the first day. Since they take 2 hours and 1 hour, respectively, the team could finish them during the 4-hour afternoon. Instead, the project manager decides to wait, allows task 2.1 to have a duration of 1 day, and starts the successor tasks on day 2. For tasks 3.1 and 2.2, the project manager puts an "x" in the day 2 column, indicating that these tasks begin (and complete) on day 2. Task 3.2 begins once tasks 3.1 and 2.2 are complete so this task begins (and finishes) on day 3. After the team submits the application for the permit, the project needs to wait while the application is processed. The town suggests that projects allow 10 days for processing of permits, so the schedule shows 10 days of delay by putting a "d" (for delay) in the boxes for days 4–13. Note that 3 of these 10 days are weekend nonworking days.

The page only has space for 14 days, so tasks that start on day 15 or later are shown in the later pages of the table, and the cells are grayed out to indicate that the information for these tasks (tasks 1.1 through 3.4) are on different pages in the schedule. Tasks 4.1 and 4.2 are listed at the bottom of the page. These are project management tasks and they run through the life of the project. The project manager allocates 8 hours for each task over the 6 weeks of the project. He decides to allow 2 hours each for weeks 1 and 2 and 1 hour each for the remaining 4 weeks of the project. Rather than allocate a small

	Applied					Day														
Task	Time	% Time	Dur	Resource	Wk Day	1	2	3	4	5			6	7	8	9	10			
					Cal Day	1	2	3	4	5	6	7	8	9	10	11	12	13	14	
2.1 Create Schematic	4 hrs		1 d	1 person		x														
3.1 Det Permits & Inspec	2 hrs		1 d	1 person			x													
2.2 Approval Schematics	1 hr		1 d	1 person			x													
3.2 Obtain Permits	2 hr		1 d	1 person				x												
				fee																
delay waiting for permits			10 d						d	d	d	d	d	d	d	d	d	d		
1.1 List Matl & Supplies	4 hr		1 d	1 person																
1.2 Comp Prices	8 hr		1 d	1 person																
delay waiting for supplies																				
1.3 Arrange Delivery or Pickup	6 hrs		1 d	1 person																
2.3 Construct Deck	40 hours	50%	10 d	1 person																
				supplies																
3.3 Schedule Inspections	3 hrs		1 d	1 person																
delay waiting inspection			3 d																	
3.4 Conduct Inspection	6 hrs		1 d	1 inspector																
4.1 PM Docs	8 hrs		proj	1 person		x	x	x	x	x			x	x	x	x	x			
4.2 PM Services	8 hrs		proj	1 person		x	x	x	x	x			x	x	x	x	x			

Figure 4.6 Schedule for deck project. (*Continued*)

	Applied																			
Task	Time	% Time	Dur	Resource	Wk Day	11	12	13	14	15			16	17	18	19	20			
					Cal Day	15	16	17	18	19	20	21	22	23	24	25	26	27	28	
2.1 Create Schematic	4 hrs		1 d	1 person																
3.1 Det Permits & Inspec	2 hrs		1 d	1 person																
2.2 Approval Schematics	1 hr		1 d	1 person																
3.2 Obtain Permits	2 hr		1 d	1 person																
				fee																
delay waiting for permits			10 days																	
1.1 List Matl & Supplies	4 hr		1 d	1 person		x														
1.2 Comp Prices	8 hr		1 d	1 person			x													
delay waiting for supplies								d	d	d										
1.3 Arrange Delivery or Pickup	6 hrs		1 d	1 person								x								
2.3 Construct Deck	40 hours	50%	10 d	1 person									x	x	x	x	x			
				supplies																
3.3 Schedule Inspections	3 hrs		1 d	1 person																
delay waiting inspection			3 d																	
3.4 Conduct Inspection	6 hrs		1 d	1 inspector																
4.1 PM Docs	8 hrs		proj	1 person		x	x	x	x	x			x	x	x	x	x			
4.2 PM Services	8 hrs		proj	1 person		x	x	x	x	x			x	x	x	x	x			

Figure 4.6 (Continued) Schedule for deck project.

(Continued)

	Applied																			
Task	Time	% Time	Dur	Resource	Wk Day	21	22	23	24	25			26	27	28	29	30			
					Cal Day	29	30	31	32	33	34	35	36	37	38	39	40			
2.1 Create Schematic	4 hrs		1 d	1 person																
3.1 Det Permits & Inspec	2 hrs		1 d	1 person																
2.2 Approval Schematics	1 hr		1 d	1 person																
3.2 Obtain Permits	2 hr		1 d	1 person																
				fee																
delay waiting for permits			10 d																	
1.1 List Matl & Supplies	4 hr		1 d	1 person																
1.2 Comp Prices	8 hr		1 d	1 person																
delay waiting for supplies																				
1.3 Arrange Delivery or Pickup	6 hrs		1 d	1 person																
2.3 Construct Deck	40 hours	50%	10 d	1 person		x	x	x	x	x										
				supplies																
3.3 Schedule Inspections	3 hrs		1 d	1 inspector									x							
delay waiting inspection			3 d											d	d	d				
3.4 Conduct Inspection	6 hrs		1 d	1 person													x			
4.1 PM Docs	8 hrs		proj	1 person		x	x	x	x	x			x	x	x	x	x			
4.2 PM Services	8 hrs		proj	1 person		x	x	x	x	x			x	x	x	x	x			

Figure 4.6 (Continued) Schedule for deck project.

amount of time to each day in the week, he records the week's project management hours for Monday but understands that this time will be consumed over the week. The budget (Figure 5.1) illustrates this allocation of time to the first day of each week. This is 16 hours for project management for a project with about 76 hours of applied time; 17% of total project time of 92 hours for project management may be a bit high but certainly within reason.

Similar logic follows for pages 2 and 3 of the table. Task 1.3 (Arrange Delivery or Pickup) is scheduled for a Sunday. Normally, the project manager would schedule this task for the next work day (not a weekend day), but he decides that it would be good to have the materials available so that the team can begin deck construction on Monday. These are decisions that the project manager makes to trade off the inconvenience of working on weekends versus completing the project earlier. Also note that for task 2.3, the resource is only working half-time, so only 4 hours per day. At this rate, the 40 hours for deck completion takes 10 working days plus 2 weekend days.

Based on this schedule, the project manager plans the first deck party 41 days after the team starts working, assuming no delays.

The project manager compares his calculated schedule with his original estimate in the project charter. The original estimate and the detailed schedule both indicate project completion in about 6 weeks or 30 working days.

4.7 Determine Schedule

To calculate the project schedule, the project manager takes the following steps:

- List all the paths through the network diagram.
- For each path
  - Add up the durations of all the tasks.
  - Add in any weekend or other nonwork time.
  - Add in any delays between tasks. For example, there might be a delay between the time that materials are ordered and when they are available in the store for pick up or delivery.

- The longest path (the path with the largest number) is the shortest time in which the work of the project can be completed. All of the other paths have "free time" (much like Paul in the reunion example above).
- Compare the estimated completion date with the completion date in the project charter. If there is significant variance, the project manager reassesses whether this project is still a viable idea.

For the deck project, these steps are as follows. The project manager uses durations and not applied time to determine project schedule. Each start to finish sequence represents a path through the network diagram.

Schedule network diagram without nonworking time:

Start → 2.1 (1 day) → 3.1 (1 day) → 3.2 (1 day + 10 days delay) → 1.1 (1 day) → 1.2 (1 day + 3 days delay) → 1.3 (1 day) → 2.3 (10 days) → 3.3 (1 day + 3 days delay) → 3.4 (1 day) → Finish–yielding 34 days of work and delay plus 10 weekend days minus 4 days of work done on weekends = 40 days.

Start → 2.1 (1 day) → 2.2 (1 day) → 3.2 (1 day + 10 days delay) → 1.1 (1 day) → 1.2 (1 day + 3 days delay) → 1.3 (1 day) → 2.3 (10 days) → 3.3 (1 day + 3 days delay) → 3.4 (1 day) → Finish–yielding 34 days of work and delay plus 10 weekend days minus 4 days of work done on weekends = 40 days.

Start → 4.1 (1 day) → Finish.

Start → 4.2 (1 day) → Finish.

4.7.1 Checkpoint

Using the schedule for the deck project (above) and the Community Gardens project schedule in Appendix G as examples, create a chart (similar to Figure 4.6) for your project schedule from the list of tasks in your schedule network diagram. Add durations, percent work time, and resources to your project schedule chart. Be sure to include task durations and any delays (or wait time) between tasks. You may find it useful to determine all paths from start to finish and calculate project completion date based on work time, delay time, nonworking time, and any work completed during nonworking time, as illustrated above for the deck project.

Compare the project schedule with the project charter, the WBS and any other project plan documents created earlier, and make any necessary changes to the WBS or the plan. Make sure that the project completion date is in line with the project completion dates in the project charter and project scope statement. Finally, get the appropriate stakeholders to sign off on this work.

5
Project Budget

Projects cost money; the key stakeholders expect that the product of the project will generate more benefit (in the form of money/revenue, enjoyment, satisfaction, or other intangibles) than it costs. In the case of the deck project, the property owner expects that the additional rent that he may obtain in the future is greater than the cost of the deck. The tenants, who may help build the deck, expect that the enjoyment of using the deck while they rent the house is greater than the sweat equity that they will contribute toward the building of the deck. However, in both cases, the return on the investment occurs after the project is complete. The sponsor and customers expend time and money in expectation of a future reward. In addition to calculating the cost of the project and the expected return of the project, the project manager and key stakeholders consider the amount and timing of investment necessary to fund the project.

To determine project costs, the project manager revisits the project schedule. For each task, he calculates the labor costs as the hours multiplied by the hourly rate. He estimates hourly rates by asking a local contractor or vendor what they charge, asking others what they have recently paid for this type of service, or using publicly available data on cost per hour (or other unit of measure) for a variety of services. He lists, for each task, any materials or services that he plans to purchase. The aggregate of these costs is the projected cost or project budget. In some situations, he adds a small amount (maybe 10% of the project budget) as a reserve to protect against cost overruns. If the project is lengthy, he can use the schedule to aggregate costs by month or quarter and monitor the actual spending against plan. He uses this as an early warning against overspending and addresses overspending before it becomes a big problem. He also monitors use of the reserve fund. This is money set aside as insurance against unexpected or low probability expenses. If he is 10% through the project schedule, and he has used 90% of the reserve fund, he likely has a spending problem; the project will go over budget unless he does something to fix the overspending.

5.1 Accounting for Apparently Free Labor Resources

In some projects, the team contributes labor; if the project manager doesn't account for this apparently free labor, he is underestimating the cost of the project. For example, when planning a wedding, the team doesn't usually charge for their time. If they hired a wedding planner, he would charge the project either by the hour or a flat rate for the entire project. Some people enjoy quilting; the cost of the quilt is quite reasonable if the project manager only calculates the material costs. If he adds in the many hours of labor that quilters work on the quilt, it would be many times more expensive. In the case study, the project manager does not calculate the cost of the labor contributed to the Community Gardens.

Sometimes, the project manager values the product of the project based on the total cost (labor plus materials). For example, if he plans to sell the quilt, he might price it based on the hours to create it or set the price roughly equal to the price of other similar quilts on the market. When someone owns his home, he doesn't count the cost of mowing his own lawn into his monthly expenses. But, if he gets a wonderful job offer out of town, he may choose to rent the house rather than sell it. He needs to charge enough rent to cover all expenses, including the cost of now hiring someone to mow the lawn (unless he requires that the tenant do this work in consideration for a lower rent).

In the business world, companies often pay people a fixed salary in exchange for a unspecified number of work hours each week. Since the people are already on payroll, they don't consider the costs when these people work on projects in addition to their regular work. If the people are already working full time on their "regular job," then they work extra hours to accomplish the work of the project. The project time may be free to the company, but it costs the employees in the form of extra time at work. Alternatively, if the people don't need 40 hours of work to accomplish their "regular" job, they may use the extra time to work on project work. In this case, the business overpays for the regular work (they planned on paying $1000 for 40 hours of work ($25 per hour), and they are paying $1000 for 30 hours of work ($33 per hour). The better way to account for this is to charge 30 hours of work for the regular work ($25 × 30 hours) and charge the project for the time spent working on the project ($25 × 10 hours).

In our examples of the deck and the Community Gardens, there is no direct charge for the volunteer labor. We capture hours worked and could assign an hourly rate to calculate a labor cost for the project.

5.2 Project Funding

The project manager also considers how to get the money to fund the project costs (labor or material or both). In the example of the deck, the tenant purchases the materials and asks for reimbursement from the landlord. The gardeners front the money for the Community Gardens and may be reimbursed when they sell the produce at the farmers market.

5.3 Budget for the Deck Project

The budget for the deck might look as follows:

Labor Hours—92 hours total [16 hours for project management, 76 hours for product work]. This is the sum of applied time listed on both Figures 4.6 and 5.1. To derive a cost, the project manager uses an hourly rate ($40) and multiplies by number of hours. For greater accuracy, he might use different rates for different types of resources.

Materials/Supplies—Total = $2925 [Permits—$125, Deck Materials—$2500, Inspection—$300]. Some of these costs are easily determined by calling the vendor or looking up fees on the vendor website. For example, the fees for permits and inspections are easily obtained from the town website. Sometimes these fees are listed on the application.

There are a number of ways to determine the cost of the materials for the deck. One approach is to determine the cost of other similar projects and increase or decrease the cost based on the relative complexity of the current project. For example, the project manager may recall that a neighbor on Maple Street built a deck a couple of years ago, but

	Applied					Day														
Task	Time	% Time	Durat	Resource	Wk Day	1	2	3	4	5			6	7	8	9	10			
					Cal Day	1	2	3	4	5	6	7	8	9	10	11	12	13	14	
2.1 Create Schematic	4 hrs		1 d	1 person		4														
3.1 Det Permits & Inspec	2 hrs		1 d	1 person			2													
2.2 Approval Schematics	1 hr		1 d	1 person			1													
3.2 Obtain Permits	2 hr		1 d	1 person				2												
				Fee				125												
delay waiting for permits			10 d						d	d	d	d	d	d	d	d	d	d		
1.1 List Matl & Supplies	4 hr		1 d	1 person																
1.2 Comp Prices	8 hr		1 d	1 person																
delay waiting for supplies																				
1.3 Arrange Delivery or Pickup	6 hrs		1 d	1 person																
2.3 Construct Deck	40 hours	50%	10 d	1 person																
3.3 Schedule Inspections	3 hrs		1 d	1 person																
				supplies																
delay waiting inspection			5 d																	
3.4 Conduct Inspection	6 hrs		1 d	1 inspector																
4.1 PM Docs	8 hrs		proj	1 person		2	x	x	x	x			2	x	x	x	x			
4.2 PM Services	8 hrs		proj	1 person		2	x	x	x	x			2	x	x	x	x			

Figure 5.1 Budget for deck project. *(Continued)*

Task	App Tim	% Time	Durat	Resource	Wk Day	11	12	13	14	15			16	17	18	19	20			
					Cal Day	15	16	17	18	19	20	21	22	23	24	25	26	27	28	
2.1 Create Schematic	4 hrs		1 d	1 person																
3.1 Det Permits & Inspec	2 hrs		1 d	1 person																
2.2 Approval Schematics	1 hr		1 d	1 person																
3.2 Obtain Permits	2 hr		1 d	1 person																
				fee																
delay waiting for permits			10 days																	
1.1 List Matl & Supplies	4 hr		1 d	1 person		4														
1.2 Comp Prices	8 hr		1 d	1 person			8													
delay waiting for supplies								d	d	d										
1.3 Arrange Delivery or Pickup	6 hrs		1 d	1 person								6								
2.3 Construct Deck	40 hours	50%	10 d	1 person									4	4	4	4	4			
				supplies									2500							
3.3 Schedule Inspections	3 hrs		1 d	1 person																
delay waiting inspection			5 d																	
3.4 Conduct Inspection	6 hrs		1 d	1 inspector																
4.1 PM Docs	8 hrs		proj	1 person		1	x	x	x	x			1	x	x	x	x			
4.2 PM Services	8 hrs		proj	1 person		1	x	x	x	x			1	x	x	x	x			

Figure 5.1 (Continued) Budget for deck project.

(Continued)

Task	App Tim	% Time	Durat	Resource	Wk Day	21	22	23	24	25			26	27	28	29	30			
					Cal Day	29	30	31	32	33	34	35	36	37	38	39	40			
2.1 Create Schematic	4 hrs		1 d	1 person																
3.1 Det Permits & Inspec	2 hrs		1 d	1 person																
2.2 Approval Schematics	1 hr		1 d	1 person																
3.2 Obtain Permits	2 hr		1 d	1 person																
				Fee																
delay waiting for permits			10 d																	
1.1 List Matl & Supplies	4 hr		1 d	1 person																
1.2 Comp Prices	8 hr		1 d	1 person																
delay waiting for supplies																				
1.3 Arrange Delivery or Pickup	6 hrs		1 d	1 person																
2.3 Construct Deck	40 hours	50%	10 d	1 person		4	4	4	4	4										
				supplies																
3.3 Schedule Inspections	3 hrs		1 d	1 person									3							
				Fee									300							
delay waiting inspection			5 d											d	d	d				
3.4 Conduct Inspection	6 hrs		1 d	1 inspector													6			
4.1 PM Docs	8 hrs		proj	1 person		1	x	x	x	x			1	x	x	x	x			
4.2 PM Services	8 hrs		proj	1 person		1	x	x	x	x			1	x	x	x	x			

Figure 5.1 (Continued) Budget for deck project.

that deck was twice the size of the planned deck. He might find out what they paid for materials and decrease that amount by an appropriate amount. He might halve the cost because his deck is smaller but increase it slightly to allow for the current higher cost of building supplies. If he plans on using higher grade materials (fancy wood for example), he would factor this into the estimate. This is generally the approach taken when developing the charter; this strategy is called analogous estimating.

A more detailed approach would be to determine how much wood is needed, the type and number of nails, fasteners, and other disposable supplies. The project manager adds up the cost of these materials to obtain a very accurate cost for the deck materials. This estimating approach tends to take a lot of time (relative to the other approaches); the project manager generally uses it as he finalizes the budget at the end of the planning. This estimating approach is called bottom-up estimating.

A middle ground is to determine the cost based on some metric. For example, there are data on the Internet that suggests costs for a deck of specified square footage and grade of materials. There are per square foot cost ranges for a wood deck with variants for grade of materials and whether it is installed by a homeowner or built by a contractor. This estimating approach is called parametric estimating; this is how the project manager derived the $2500 figure used above.

The project manager further decomposes the budget into weekly amounts that aggregate labor hours and material costs:

Week 1—13 hours, $125 permits
Week 2—4 hours, no material costs. Note that Week 2 is mostly delay waiting for permits to be approved
Week 3—14 hours, no material costs. There is some delay waiting for supplies
Week 4—28 hours, $2500 costs for deck materials. The 28 hours includes the 6 hours on Sunday, day 21
Week 5—22 hours, no material costs
Week 6—11 hours, $300 inspection fee

As with the project schedule, the project manager compares the cost (92 labor hours and $2925 for materials) against his original estimate documented in project charter ($3000 for materials) and determines that the project costs are reasonable relative to estimates in the charter. If the charter contains labor estimates (in hours or dollars), the project manager compares the labor estimates in the budget to the original estimates in the charter.

The deck project budget in Figure 5.1 documents these costs.

Appendix H contains the project budget for the Community Gardens project.

5.3.1 Checkpoint

Now it is time for you to calculate the costs for your project. For each task on your schedule, replace the "x" with the hours allocated for that task. On the line below, enter any costs for materials or supplies. Sum these values for overall project costs. Sum the values for each week or month to develop cost estimates for each project week, month, or quarter. Then compare this budget with your budget estimate in your project charter. If there is significant variance, talk with your sponsor and other key stakeholders and decide whether to continue with the project; determine if changes are necessary to scope, schedule, budget, and quality to align stakeholder expectations, project commitments, and available funding.

6
Project Quality

Quality is a desired characteristic of the products and services we acquire; it often means different things to different stakeholders. For example, when buying a new car, quality may mean luxury in the form of heated leather seats and adaptive cruise control to one buyer; to another buyer, quality may mean reliability and economical vehicle operation and maintenance. Sometimes stakeholders are able to articulate the characteristics of quality that they value; other times they cannot put their quality requirements into words and they argue that they know quality when they see or experience it.

Quality is a characteristic of a project or product; quality projects or products are those that meet stakeholder expectations—there is good quality and poor quality.

6.1 Quality Project versus Quality Product

Sometimes stakeholders characterize quality in terms of the product of the project; other times, they characterize quality in terms of the process that generates a quality product. Project processes such as charter, scope statements, and other documents and tactics addressed in this book are examples of project quality that incents product quality. A quality process generates a product or service that consistently meets the stated requirements.

Franchises like McDonald's and Burger King promise predictability of their products. Wherever a stakeholder orders fries or a milkshake or burger, the product is the same. The parent company prescribes a process, including equipment, raw materials, and steps to create and cook the company products. When stakeholders cook similar things at home, they generally don't specify a process; they may buy the raw materials at different stores depending on where there is a sale and may use an outdoor grill or indoor broiler depending on the season or the weather. Cooking at home, we think in terms of the product (a medium-sized hamburger, well done, on a roll), rather than a predictable and repeatable process. In the home environment, the customer requirements are implicit, and the process tends to be predictable because the few people doing the grilling have cooked for the same stakeholders for quite some time. In the franchise environment, there are hundreds of people doing the grilling and thousands of customers with expectations. In the franchise environment, consistent and predictable products depend on a well-defined and communicated process.

Stakeholders define consistent and predictable products with metrics which specify the required or desired characteristics of the project and product. The quality plan specifies these project and product metrics.

6.2 Subjective versus Measureable

Sometimes the characterization of quality is subjective; other times, quality is measurable. For example, a quality measure for the deck might be that it meets local building codes. The team researches the local building codes and specifies the various characteristics in terms of materials, thickness, or weight-bearing capability of the deck floor and any required railings or other safety features.

Subjective measures of quality may take the form of a requirement; for example, the property owner might specify that the deck fits the "look and feel" of the building. Stakeholders may also have a subjective measure for the completed deck; they may comment that "the deck looks great." Subjective measures of quality are especially challenging for project managers to successfully implement.

6.3 Quality Plan

The quality plan documents the desired quality and how to measure it. It costs time and money to ensure a quality product or project; stakeholders may not be willing to pay for quality in excess of that which is necessary. For example, it may be necessary that the materials for the deck are of sufficient quality that the finished deck meets all building codes and safety requirements. It may not be necessary to use very expensive wood that looks great but doesn't contribute to the requirements of building regulations and safety. If there is an additional requirement that the deck match the "look and feel" of the building exterior, it may be necessary to use more expensive materials to satisfy these requirements as documented in the charter and scope statement.

Scope and quality are related, and there may be a temptation to embed quality characteristics and requirements in the scope statement. The risk is that the quality characteristics get overlooked as the team focuses on the scope of the project. Setting the quality plan as a separate activity and document ensures that quality does not get overlooked as the team focuses on the project and product scope.

It is not always necessary to require 100% quality; it may be too expensive in terms of time and money or it simply may not be appropriate to the current project.

The quality plan allows for a prioritization of the project core components of scope, schedule, and budget as high, medium, or low. This allows the key stakeholders and project team members to choose among alternatives if there are conflicts or significant variances. For example, if scope is most important, the team understands that it is preferable to take longer to complete the project or add more resources (and therefore increasing costs) rather than eliminate any product features.

For both the project and the product of the project, the quality plan includes a categorization, a description of the quality metric, a target metric, the allowable variance, and actual results collected after the project completes:

- *Priority*: The project manager works with the sponsor and customer to prioritize project scope, schedule, and budget. The relative priority of these core project components drives trade-offs among project alternatives.
- *Category*: Project metrics generally follow the core project components of scope, schedule, and budget. Prioritizing these core project components and specifying acceptable performance is a basic project management best practice. At a minimum, project success and project management performance are measured by delivering project scope and product features on time and within the project budget.

  Product metric categories are a function of the product of the project. For example, the product categories for the deck project are very different from the product categories for the Community Gardens. In the case of the deck project, the team is focused on safety, regulations, aesthetics, and building materials characteristics such as water resistance. In the case of the Community Gardens, the team is focused on the infrastructure of the garden plots, the appearance and quantity of the produce, and their success in the 4H competition.

Organizing the project and product metrics into categories aligns the project metrics with the core project components and aligns the product metrics with the top level WBS categories or other key product attributes. It is easier for the project team to identify target metrics by category than it is to identify target metrics thinking of the project overall. As with other brainstorming and decomposition activities such as the WBS, the team starts with the top level project and then moves to smaller, more specific components.

- *Description*: An explanation of what will be measured, the description should be sufficiently concrete that the team creates measureable metrics. For project metrics, the quality plan describes project capabilities; for the product metrics, the quality plan describes product features. The project metrics for the deck include the core project components of scope, schedule, and budget. For some projects, the project manager may include other project processes such as risk. He might be concerned that the team captures the majority of project risks and might include a risk metric such as no more than five unanticipated risks occur.

Occasionally, the quality measure is not objective and measureable. For the deck project, it is difficult to quantitatively assess whether the project meets the purpose as defined in the charter. The purpose in the charter stated "create an outdoor space to enjoy family activities during nice weather." The project manager addresses the difficulty of quantitative assessment by using sign off and approval by the project sponsor and project customer as satisfaction of the metric. The property owner and the renters are both sponsors and customers. They are both sponsors, in that the property owner contributed money and the renters contribute resources to build the deck. They are both customers, with the property owner being a longer-term customer than the renter. For the deck product metric, the aesthetics category is subjective as it is difficult to objectively assess whether the deck complements the look and feel of the house; the project manager still creates a metric in the form of satisfaction of the owner and the renters.

- *Target Metric*: The target metric identifies how the project manager determines if the project or product satisfies the description.
- *Allowable Variance*: This is based on the target metric. Working with key stakeholders, the project manager determines how much variance is acceptable.
- *Actual Performance*: The project manager monitors actual performance relative to the plan during the project execution. If there are variances in excess of the allowable amount, the project manager either makes changes to the project execution to bring the project within the allowable variance or initiates a change to the project plan. After the project completes, the project manager completes the last column of the quality plan and identifies what worked well and what didn't work during the project lessons learned.

Table 6.1 shows the quality plan for the deck project.

Appendix I contains the quality plan for the Community Gardens project.

6.3.1 Checkpoint

Now it's your turn to identify what makes your project and the product of your project quality deliverables. Determine the appropriate quality priorities, categories, and capabilities (project) or features (product). For each capability and feature, identify target metrics and allowable variances. Based on this quality plan, determine if you need to make any changes to your project plan including scope, schedule, or budget. If so, revise these documents now. Have the appropriate stakeholders sign off on the changes and approve the newly created quality plan.

6.4 Next Steps

At this point, you have a good idea of the "what"—what the project is and what it will accomplish. This is the scope and quality. And you've identified some constraints in the form of schedule and budget. To make sure these core project components are complete, the project team spends time determining what might go wrong and how they might capitalize on project opportunities. These risks are identified in the next chapter.

Table 6.1 Quality Plan for Deck Project

Priority [1 = high, 2 = medium, 3 = low]: 1-Scope; 3-Schedule; 2-Budget				
PROJECT METRICS				
CATEGORY	DESCRIPTION—PROJECT CAPABILITIES	TARGET METRIC	ALLOWABLE VARIANCE	ACTUAL PERFORMANCE
Scope	Project meets purpose as defined in charter	Sign off and approval by sponsor and customer (property owner and renters)	None	
Schedule	Project meets schedule constraints	Schedule milestones as defined in project schedule	±10% or as defined in schedule	
Budget	Project meets budget constraints	Costs incurred as defined in project budget	±10% or as defined in budget	
PRODUCT METRICS				
CATEGORY	DESCRIPTION—PRODUCT FEATURES	TARGET METRIC	ALLOWABLE VARIANCE	ACTUAL PERFORMANCE
Regulatory	Deck meets all building codes	Obtain building permit	No acceptable variance	
Safety	Deck supports expected load of people, equipment & furniture	Supports combined weight of 2000 pounds	±10% metric acceptable	
Aesthetics	Deck complements look and feel of house	Owner satisfied Renters satisfied	Not applicable	
Usage	Deck in use for at least 3 seasons	Renters report using deck April through October	Renters report actual use limited to May through Sept.	

7
Project Risks

At this point, the project has a defined and approved scope, schedule, budget, and quality; the project manager now focuses on creating an environment for project success. There is always a possibility that things might go wrong and get in the way of meeting the project scope, schedule, budget, and quality goals. Often, there are opportunities to exceed these project goals. The project manager works with his team to identify all the threats to the project and all the opportunities to exceed the project goals. He neutralizes the most significant threats to the project and encourages opportunities. For the most significant threats and opportunities, he spends time as part of the planning process to *proactively decide what he will do if these risks occur* at a future point in time; these are "Risk Response Plans."

These plans document what the project manager will do if these threats or opportunities to the project occur, and what he will do to increase the probability that opportunities are realized and decrease the probability that threats occur. He adds the work of the risk response plans to the project WBS and adds the necessary time to the project schedule and money to the project budget. If the risk does occur (and since these are the most significant risks, it is likely that at least some of these risks will materialize), he already has time and money in the project plan, and he has defined what needs to be done. And for those risks that don't occur, he can use the schedule and budget for other risks or bring the project in ahead of schedule and under budget—a very good thing for a project and project manager!

A risk is a future event; something that has not yet occurred. Threats are negative risks, opportunities are positive risks. Stakeholders generally think of risks as negative events, where bad things happen on a project. Opportunities (or positive risks) are unexpected good things that may happen to a project.

As an example of a positive risk in building the deck, the team finds wood planks that are pretreated such that they don't require any weatherproofing or painting, finishing, or varnishing. They cost a bit more, but the team won't have to purchase weatherproofing and paint, and will save time by not having to treat the deck once finished. The project saves both time and money. The team has an *opportunity* to have a positive impact on schedule and budget.

To manage project risk, the project manager does the following:

- Creates a list of risk categories
- Creates a list of risks
- Prioritizes the risks
- Creates risk response plans

Once he completes the risk response plans, he revisits the WBS, schedule, budget, and quality and makes changes to the project plan to include this additional work.

7.1 Naming the Risks

When challenged to create a list of risks, team members often come up with one word risk descriptions such as weather, schedule, budget, and quality. Unfortunately, this one word characterization of risk doesn't capture the real risk. For example, for the deck project a stakeholder might say that weather is a risk. Does he mean that there is a risk of too much bad weather or does he mean a risk of too much good weather? And, what is the impact of too much bad weather?

A better phrasing is "There is a risk of ________________ with a consequence of ________________." This wording forces a stakeholder to state the risk separately from the consequence of the risk.

The stakeholder might rephrase the risk as follows: "There is a risk of more bad weather than anticipated with a consequence that the schedule will be delayed." This suggests that he has allowed some contingency for days when the weather precludes outside work and that the actual days of bad weather exceed that time. The consequence (schedule delay) suggests things that the project manager might do to mitigate this risk, such as having additional resources on standby or having a list of available contractors to augment his work crew.

Table 7.1 Risk Categories

PRODUCT RISKS	ENVIRONMENTAL/EXTERNAL RISKS
• Requirements/Scope • Design and technology • Quality and feasibility	• Vendors • Legal/Regulatory/Safety • Market • Customer/Stakeholder
PROJECT ENVIRONMENTAL RISKS	PROJECT MANAGEMENT RISKS
• Decision making • Change management • Life cycle • Conflict management • Organization structure/ Dysfunction	• Executive/Sponsor • Scope • Schedule • Budget/Cost • Quality • Resources

7.2 Risk Categories

Risk categories facilitate the generation of a complete list of risks and provide a structure for organizing the team's thoughts about risks. Some likely risk categories might be project management risks, product-specific risks, environmental risks that are external to the project, and project environmental risks, similar to those described in the chapter on project toolkit. These categories are then further decomposed into more specific risk categories. Table 7.1 lists possible risk categories.

7.3 List of Risks for the Deck Project

To create a list of risks for the deck project, the project manager meets with his team to brainstorm negative and positive risks. To engage the team in the process, the project manager asks the team members to write risks on sticky notes. The project manager later combines sticky notes to eliminate duplicates and organizes the sticky notes into the risk categories. As team members often focus on product risks rather than project risks (usually because the product is tangible and easier to think about), the project manager should separately create lists of project risks and product risks.

Table 7.2 contains a list of the risks for the deck project.

Appendix J contains the positive and negative risks for the Community Gardens project, along with a table of risk categories.

Table 7.2 Risks for Deck Project

NEGATIVE RISKS (THREATS)	POSITIVE RISKS (OPPORTUNITIES)
• There is a risk that the neighbors object (due to noise) and they block the permit with the consequence that the deck cannot be built. (stakeholder risk) • There is a risk that PM inexperience leads to quality problems with a consequence that the deck doesn't pass inspection and it cannot be used. (quality risk) • There is a risk that the soil cannot support the weight of the deck and the team needs to reinforce the soil with a consequence that it takes longer and costs more. (scope, schedule, and cost risks) • There is a risk that the wood cannot handle the weight and it sags and splinters with a consequence that we need spend more money to reinforce and refinish the deck. (quality, cost risks)	• The small interested group of stakeholders provides an opportunity to get accurate scope and timely approvals. (stakeholder risk) • Neighbors decide they want to build decks also; we have an opportunity to buy materials in bulk with a consequence that we save money. (cost risk) • The team discovers a synthetic wood product that lasts much longer than more traditional wood deck products with a consequence that the life of the deck is much longer. (scope and quality risk) • The young children in the neighborhood like playing on the deck; as a consequence, we have an opportunity to monitor their activities. (safety risk)

7.4 Checkpoint

Consider the risks to your project and create a list. Try using the phrasing suggested above and create separate lists for positive risks and negative risks. Using the risk categories from Table 7.1, record the risk category for each item. You may find that you need additional risk categories—this is fine and will result in an expanded list of risk categories that you can use for future projects. Use the categories to jog your thinking to add additional risks to each category. Working with your team, the sequence of actions might:

- Create a list of risk categories.
- Create a list of risks.
- Sort the risks into the categories.
- Think about each category; decide if there are other risks specific to the risk category.
- If there are risks that do not fit into an existing category, create additional risk categories as necessary. Keep in mind that you can reuse the risk categories from one project to the next.

7.5 Probability and Impact

On a complex project, the project manager may have a long list of negative and positive risks; it is often not practical to address all of these risks. A best practice is to prioritize the negative risks and the positive risks and develop plans to mitigate the most significant negative risks and optimize the most significant positive risks. Rather than develop these plans for all the negative and positive risks, the project manager develops plans only for the risks that will help the most (positive risks) or hurt the most (negative risks) if they are realized.

A widely used approach is to develop a risk score and compare the risk scores for all risks, or for all risks within each risk category. The risk score is a blend of probability, impact, and urgency.

Probability is the likelihood that an event will occur; impact is the effect or influence of the event.

For example, what is the probability that the project manager will be struck by lightning within the next 60 seconds? Unless he is outdoors in the middle of a thunderstorm holding up a metal golf club, the likelihood is small. But, if he is struck by lightning, the impact is likely to be very big.

Urgency suggests that a "risk event" may be imminent. For example, a weather *watch* for lightening suggests conditions that lightening may occur in the next several hours. A weather *warning* for lightening suggests that lightening is imminent and you should take shelter immediately. Another consideration is whether the project manager is in an area that is prone to this type of risk. For example, tornadoes often occur in areas called "tornado alley" so he needs to consider if he is in an "environment" where the risk event is likely to occur.

Still another consideration might be the expertise of the source of the risk assessment. A weather warning from NOAA (National Oceanic and Atmospheric Administration) or a weather station has more credibility than a neighbor predicting a storm based on his sciatica pain.

7.6 Risk Score

The risk score is the probability times the impact (and times urgency if appropriate). The project manager is not creating absolute values

Table 7.3 Probability and Impact Scales

PROBABILITY		IMPACT	
Very high	10	High	10
High	6	Medium	5
Low	4	Low	1
Very low	1		

for probability and for impact for each listed risk. Rather he is developing a score so he can rank order the risks relative to each other and select the top few for further analysis and development of a risk response plan. To create a ranking for probability and impact, he uses descriptive ranks and assigns numeric values to each. For example, he might use rankings of high, medium, or low, or another ranking might be very high, high, low, very low. Then he assigns numeric values to each; he might use a scale of 1–10 or 1–100. It doesn't matter as long as he is consistent and the scale is easy to use. So, for example, he might use a scale of 1–10 and organize it as illustrated in Table 7.3.

If a risk has a high probability and a medium impact, it has a risk score of 30 (probability = 6 and impact = 5). If it has a very low probability, and a high impact, it has a risk score of 10 (probability of 1 and impact of 10). In this example, the first risk is ranked higher than the second risk. The rankings do not need to be precise, it is sufficient to be consistent and that the rankings are approximately correct. If the project manager wishes to further weight the risk score taking urgency into account, he might assign a value of 1 if not urgent, value of 2 if moderately urgent, value of 3 of very urgent. If adding values for urgency, he might want to use a scale of 1–50 or 1–100 for probability and impact so the urgency scores do not overwhelm the scores for probability and impact.

Depending on the project, the project manager may wish to ignore the numeric weights for probability and impact and simply assign risk scores as shown in Table 7.4.

In these cases, he may eliminate groups with similar scores (high probability-low impact grouped with low probability-high impact, high probability-medium impact and medium probability-high impact, and medium probability-low impact and low probability-medium impact). He will have 6 groups instead of 9.

Table 7.4 Probability, Impact, and Risk Score

PROBABILITY IMPACT DESCRIPTION		
PROBABILITY	IMPACT	RISK SCORE
High	High	6
High	Medium	5
High	Low	4
Medium	High	5
Medium	Medium	3
Medium	Low	2
Low	High	4
Low	Medium	2
Low	Low	1

Once the project manager develops scales for probability and impact, he saves them for use on future projects.

There are a number of approaches to determine the most significant risks. The project manager might decide to take the top 10% of the risks for further action (in the form of a risk response plan). He might decide that any risk with a risk score greater than a specified number warrants a risk response plan. The risks that do not warrant a risk response plan stay on a risk watch list; the team monitors this list periodically to see if the risk has become more significant. The project manager adds new risks to the list as he discovers them.

These risk scores enable a discussion of the relative importance of risks based on objective criteria. This is very different from a discussion that focuses on which risk is more significant based on stakeholder perception or bias. Absent some objective criteria for assigning values and calculating a risk score, the discussion becomes very subjective.

It is a best practice to use the project team to generate the list of risks and then subsequently assign probability and impact scores. This might include the project manager, project sponsor, and key customers. As a general rule, stakeholders are more vested in a project when they have greater involvement. They tend to feel ownership of a project when they are involved in the development and execution of the plan, and as a result, work harder to make the project a success.

Table 7.5 illustrates the risk rankings and risk scores for the deck project. Table 7.2 provides the information for the first two columns of Table 7.5.

Table 7.5 Risk Scores for Deck Project

NEGATIVE RISKS (THREATS)	CONSEQUENCE	RS	PROB	IMPACT	CATEGORY
Neighbors object; concerned with noise (project risk)	Neighbors block permit; the team cannot build the deck	10	V low-1	High-10	Stakeholder
Construction inexperience (project risk)	Quality problems; deck doesn't pass inspection	30	High-6	High-5	Quality, regulatory
Soil can't support weight of deck (product risk)	Need to reinforce soil; need more time and money	20	Low-4	Med-5	Scope, schedule, cost
Wood sags and splinters (product risk)	Need more money to reinforce and refinish deck	5	V low-1	Med-5	Quality, cost
POSITIVE RISKS (OPPORTUNITIES)	CONSEQUENCE	RS	PROB	IMPACT	CATEGORY
Small, interested group of stakeholders (project risk)	Obtain accurate scope and timely approvals; stay on schedule	60	High-6	High-10	Stakeholder
Neighbors also build decks (project risk)	Buy materials in bulk and save money	20	Low-4	Med-5	Cost
Discover synthetic wood product with long life (product risk)	Deck lasts much longer than originally expected	20	Low-4	Med-5	Scope, Quality
Young children like playing on deck (product risk)	Opportunity to monitor their activities	30	High-6	Med-5	Safety

On the basis of these rankings, the order of the risks would be as follows:

Order of negative risks:

- Construction inexperience resulting in quality problems—Risk Score (RS) = 30.
- Soil cannot support weight of deck; need more time/money—Risk Score = 20.
- Neighbors object and block permit—Risk Score = 10.
- Wood sags/splinters; need more money—Risk Score = 5.

Order of positive risks:

- Small interested group of stakeholders resulting in accurate scope and approvals—Risk Score = 60.

- Young children like playing on deck; opportunity to monitor activities—Risk Score = 30.
- Neighbors also build decks; buy materials in bulk and save money—Risk Score = 20.
- Discover synthetic wood product with long life—Risk Score = 20.

On the basis of these risk scores, the project manager would likely develop risk response plans for the Top 2 positive and negative risks. In section "Risk Response Strategies," Figures 7.1 and 7.2 illustrate risk response plans for the highest ranked negative and positive risks; the remaining risk response plans are exercises for the reader.

Appendix K contains the probability and impact scales, along with the risk rankings and risk scores for the Community Gardens project.

7.7 Checkpoint

Develop a ranking system for probability and impact for your projects. Decide how many of the top risks warrant the additional effort of a risk response plan. Take the list of risks that you created and, working with your team, assign probability and impact rankings and generate a risk score. Rank order the risks and identify the top risks which warrant the additional work of a risk response plan. There is no magic number as to how many risks are top rated. It depends on the size of your project. For a large, lengthy, expensive project, you will likely have more risks identified as significant. For a simpler project, the list will be smaller. Best practice suggests that even for a simple project, you likely have 5–10 risks (some opportunities and some threats).

7.8 Risk Response Plans

For the risks that have the most potential impact on the project, the project manager *proactively* decides what he will do *if the risk occurs.* He creates risk response plans that are in place and ready to execute, if and when the risk materializes.

This is similar to the approach taken by federal agencies such as FEMA (Federal Emergency Management Agency). They are responsible for the coordination of responses to natural and man-made disasters. They don't wait until a situation occurs and then convene a group of people to determine what needs to happen. They have complete plans, including scope, schedule, budget, and resources, on the shelf; on notification that an emergency has occurred (or in some cases, is about to occur), everyone on the team goes into action as specified in the plan.

Other examples of risk response plans likely on the shelf include the following:

- Nuclear facilities have risk response plans in case of various types of failures.
- Airlines and air traffic control have risk response plans for exception situations such as weather, accidents, and security threats.
- Companies such as Amazon and eBay have plans in place in case of server failures.
- Hospitals have plans in place in case of power failure or large-scale disaster resulting in a large influx of patients.

Since risk response plans are integrated into the project WBS, the project manager needs the same information for the risk response plans as he identified for the work packages. While the work packages are part of the project, the risk response plans are only executed on a contingency basis. The project manager only implements the risk response plans if an opportunity or a threat materializes. In addition to the information collected for work packages, he provides additional information about when to implement a risk response and action to take if the risk response plan doesn't address the risk as expected.

As a general rule, the project manager includes the following in the risk response plan; for smaller projects, the project manager may find that the fallback plans, residual risks, and secondary risks are not applicable:

- Risk category and owner(s)
- Risk statement and consequence including scores for probability, impact, and risk overall
- Risk response strategy (avoid/exploit, mitigate/enhance, transfer/share, or accept)
- Actions to implement risk response strategies
- Warning signs that a risk is about to occur
- Budget and schedule

- Fallback plans—what to do if the original risk response plan is not effective
- Any risks that are likely to remain after implementing the risk response plan (residual risks)
- Any risks that occur as a consequence of implementing a risk response plan (secondary risks)
- Contingency reserves—funds to cover the cost of implementing the fallback plan if necessary

For each risk response plan, the project manager makes changes to the WBS, the schedule, and the budget as necessary to include the risk response plan in the project plan. This may result in a longer schedule or an increase in the budget. While this is unfortunate, it is better to find this out early in the planning of the project when the project manager is still finalizing project scope, schedule, budget, or quality of the project.

7.9 Risk Response Strategies

There are several approaches for responding to anticipated risks; the project manager selects one of the strategies as part of his risk response plan. One approach to risks is to reduce the uncertainty around the event; to influence uncertainty, the project manager affects probability. For negative risks, he attempts to avoid the risk; for positive risks, he tries to increase the probability that the event occurs. Another approach is to focus on the impact. For negative risks, the project manager tries to minimize the impact; for positive risks, he tries to maximize the impact. Still another approach is to partner with prospective stakeholders to influence the prospective threat or opportunity. In some situations, the most appropriate risk response strategy is to accept the risk.

These risk response categories include the following:

- *Change the certainty of the risk*
    - *Avoid the risk*: For negative risks, the team avoids the risk by developing plans and/or taking actions that remove the risk from the project. In the case of the deck, the team might avoid problems with wood quality (sagging and splinters) by using another material (a synthetic wood, a

plastic material, or concrete) to construct the deck. These likely introduce other risks but would avoid problems with the wood quality.

- *Exploit the risk*: For positive risks, the team exploits the risk. The project manager increases the probability (reduces the uncertainty) that the opportunity occurs. For the risk of young children playing on the deck, the team might paint cartoon characters on the wood planks so that young children might find the deck an entertaining place to be.

- *Modify the impact of the risk*: In this strategy, the project manager tries to optimize the impact of the risk. For negative risks, he minimizes the impact; for positive risks, he maximizes the impact:
    - *Mitigate the risk*: For those risks that have a negative impact on schedule, he might have more people working on the task so there is a smaller schedule impact.
    - *Enhance the risk*: For positive risks, the team enhances the probability that neighbors will also build decks if they meet with the neighbors and share the excitement of having a deck.
- *Involve others*: In this strategy, the team partners with other prospective stakeholders to positively influence probability, impact, or both:
    - *Transfer the risk*: For negative risks, the team transfers the risk to someone else. Auto or homeowners insurance are examples of risk transfer. In these cases, the project manager pays another party to assume the financial risk if something should happen to his property. Sometimes in business, the team purchases a surety bond, so that the project will be paid money if the subcontractor fails to deliver as per the agreement. However, even though the project is compensated financially, the schedule is compromised and the team cannot recover the lost time.
    - *Share the risk*: For positive risks, the team shares the opportunity with others. An example might be inviting a neighbor to share the cost of the deck in return for the right to use it.

- *Do nothing and accept the risk*: In this strategy, the team accepts the risk and does not take any action until the risk materializes:
  - *Active acceptance*: Includes setting aside a contingency reserve in the form of money, time, or resources.
  - *Passive acceptance*: Includes documenting the risk but taking no action until it occurs.

Risk Response Plan

Risk Name: Construction Inexperience **Date:**

Risk Category: Quality, Regulatory **Risk Owner:**

Risk Statement and Consequence: There is a risk that the team's inexperience leads to quality problems with a consequence that the deck doesn't pass inspection and it cannot be used.

Risk Score = 30 Probability: High—6 Impact: High—5

[For top ranked risks, complete the sections below.]

Response Strategies (Avoid/Exploit, Mitigate/Enhance, Transfer/Share, Accept (passive/active)): By choosing the mitigate strategy, the project manager reduces the impact of the builder's inexperience on the project quality. The local big box hardware stores offer free classes in deck construction. The builder will take this class to an intermediate level and identify experts that he can call for advice and support.

Actions to Implement the Response Strategies: Register for the next available class in deck construction.

Warning Signs of a Risk Occurrence: Not applicable initially; builder will take the course if inexperienced in deck construction. If the builder has many questions after the class and during deck construction, this may be indication of potential quality problems.

Budget and Schedule: No impact on materials budget; labor hours increase by the time to complete the class. Expect no change to duration of project.

Fallback Plans: (What the team will do if risk occurs and primary response doesn't work.) If the builder takes the deck construction class and still has difficulty with construction, the project manager will hire a construction consultant to monitor the construction. This will cost $30/hour; estimate is 10 hours.

Residual Risks: (Risks that the team expects will remain after the risk response plan is implemented.) Although the construction class provides "book learning" and limited hands on experience, the builder will still be inexperienced in deck construction so there is still some limited risk of quality issues.

Secondary Risks: (Risks that occur as a consequence of implementing the risk response plan.) As a consequence of taking the deck construction course, the builder may think he has more expertise than he actually does have, taking unwise risks or charging for his time.

Contingency Reserves: (Money set aside to cover the cost of the fallback plan.) The project will set aside $300 for consulting fees as contingency reserve.

Figure 7.1 Risk response plan deck project—Negative risk.

Risk Response Plan

Risk Name: Interested Stakeholders **Date:**

Risk Category: Stakeholder **Risk Owner:**

Risk Statement and Consequence: There is a risk that the small interested group of stakeholders provides an opportunity to get accurate scope and timely approvals.

Risk Score = 60 Probability: High—6 Impact: High—10

[For top ranked risks, complete the sections below.]

Response Strategies (Avoid/Exploit, Mitigate/Enhance, Transfer/Share, Accept (passive/active)): By choosing the exploit strategy, the project manager attempts to ensure that the project realizes this opportunity. He meets with key stakeholders one-on-one and reinforces the benefits of the project.

Actions to Implement the Response Strategies: Meetings with key stakeholders.

Warning Signs of a Risk Occurrence: The project manager knows this opportunity is real when stakeholders contact the project manager and project team and inquire as to how they can help with the project.

Budget and Schedule: No budget impact; need time to meet with stakeholders.

Fallback Plans: (What the team will do if risk occurs and primary response doesn't work.) If key stakeholders won't meet with the project manager or project team, the project manager will leverage other key stakeholders and ask that they influence the reluctant stakeholders.

Residual Risks: (Risks that the team expects will remain after the risk response plan is implemented.) There is a small chance that stakeholders will not be fully committed to the project.

Secondary Risks: (Risks that occur as a consequence of implementing the risk response plan.) Stakeholders may find the contacts intrusive and back away from the project.

Contingency Reserves: (Money set aside to cover the cost of the fallback plan.) No budget impact; small increase in applied time in meetings with stakeholders.

Figure 7.2 Risk response plan deck project—Positive risk.

Figures 7.1 and 7.2 illustrate risk response plans for the most significant (highest risk score) negative risks and positive risks for the deck project.

Appendix L contains Risk Response Plans for the Community Gardens project.

7.10 Checkpoint

For your project's top ranked risks (both positive and negative) identified earlier, complete a risk response plan. Determine if you need to make changes to the WBS, the network schedule diagram, resources/

staffing, or other project elements. Obtain stakeholder approvals for these changes and for the risk response plans.

7.11 Next Steps

At this point, you have identified the project risks, categorized them, and prioritized them. For the top risks, you have created risk response plans and you have updated the list of work packages, project activities, project schedule, project budget, project scope, and quality. The next chapter addresses the project toolkit—those tactics and processes that stakeholders use to plan and execute projects.

8
PROJECT TOOLKIT

The project manager and the team now have a partially completed project plan that includes the scope, schedule, budget, quality, and risks. The project manager and the team rely on existing document templates and proven tactics and processes to facilitate the job of defining and planning the core project. The sample content for the project charter, scope statement, and risk response plans are examples of document templates. The format for the WBS, schedule network diagram, project schedule, and product budget are examples of documents, with specified content and format. The risk categories and risk probability and impact scales are examples of standards used to prioritize project risks.

These documents, standards, tactics, and processes provide a framework so that the project manager and the team don't start with a blank slate for each new project. The project manager and team have approaches that worked in the past on which to base the work of each new project.

This framework provides a set of repeatable processes. The benefits of repeatable processes that are documented and consistently implemented include the following:

- The project manager and project stakeholders don't have to learn new processes and approaches for each project or for different parts of a given project. They have a set of proven approaches and use them consistently.
- Stakeholders of the project know what to expect. They become familiar with the definitions, documents, sequence of events, tactics, and processes. They do not have to relearn this for each project or different parts of a given project.
- Practice makes perfect—project teams and stakeholders get better at planning, managing, and executing projects when they repeat this work many times. Studies suggest that mastery

> requires 10,000 hours of practice. Well-documented and consistent content, templates, tactics, and processes enable the repetition necessary for project managers and teams to excel at the project processes.

The project manager and team perfect templates and repeatable processes over time. An example might be the risk categories discussed earlier. The project manager and team might find that their projects experience risks not included in the original risk categories. They add these newly discovered risk categories to the list, and over time the list becomes more complete for the types of projects they manage. Similarly, the project manager and team might modify the content of the documents developed as part of the project core.

Table 8.1 lists the documents, standards, tactics, and processes that project managers and teams use in the planning and management of predictable projects.

The project manager uses the content of the project toolkit for both project planning and project execution. Many of the documents and standards listed are part of the project plan discussed earlier. The project manager uses the tactics and cross-project processes during planning and execution as well.

8.1 Documents

The first column of the project toolkit table contains documents completed during the project planning phase; these documents direct the efforts of the project manager and team during project planning, management, and execution. The project manager and team have a template for each document and examples of completed templates. A template is the content and format for a document, without project-specific information. Many of these documents are part of the project core and introduced in earlier chapters. Other documents are more appropriate to the project context, including stakeholders, communication, and teaming, discussed in later chapters.

8.2 Standards

The standards, listed in the second column, are discussed in the chapter on risk. The project manager may develop and use other cross-project

Table 8.1 Project Toolkit

DOCUMENTS	CROSS PROJECT STANDARDS	TACTICS	CROSS PROJECT PROCESSES
Charter	Risk categories	Lists	Change management
Scope statement	Probability & impact scales	Graphic display of information including charts, tables, matrices	Decision making
Work breakdown structure	Risk response strategies	Meetings	Conflict management
Network diagram		MBWA	Project life cycle
Schedule		Sticky notes	
Budget		Brainstorming	
List of risks			
List of ranked risks			
Risk response plans			
Stakeholder matrix			
Stakeholder engagement matrix			
Communication plan			
RAM for documents			
RAM for WBS project deliverables			
Organization chart			
Action item list (also issues list)			
Status reports			
Change requests			
Meeting agenda			

standards for his projects. Both the documents and standards are common to all projects and appropriate to a repeatable approach to all projects.

8.3 Tactics

Tactics, listed in the third column of Table 8.1, are approaches to collecting and illustrating information. A preliminary list of tactics includes lists, graphic displays, meetings, MBWA, sticky notes, and brainstorming.

8.3.1 Lists

Lists are valuable for capturing information and communicating it to project stakeholders. Unlike documents or templates which often use full sentences, lists generally use phrases rather than full sentences, so the information is more easily accessed. Alternatives to lists for the display and communication of information are document templates and graphic displays of information.

Examples of lists include:

- Action items
- Issues
- Stakeholders
- Project documents
- Risks
- Tasks
- Work packages

The project manager and the team use the lists to check off items as they complete the work. Appropriate chapters contain examples of these lists.

8.3.2 Graphic Displays of Information

While lists are good for displaying and communicating some types of information, other information is better communicated graphically. The expression is "a picture is worth a thousand words." Often, it is easier to communicate messages with a picture than it is to communicate the same messages in text or as part of a list.

While it may be easier for the consumer of the information to understand information displayed graphically, it is often more time consuming to create a graphic presentation of information than to present the same information in words. Graphic displays are concise requiring that the project manager be clear about his message.

The project manager may have goals in mind for the Community Gardens cookbook sales. He creates a target number of cookbooks to be sold each month and generates a target monthly and cumulative income (based on a cookbook price of $10) to be used to support the gardeners and the soup kitchen. He tracks and reports the Community

Table 8.2 Community Gardens Cookbook Sales Data

CG COOKBOOK—SALES DATA (NUMBER OF BOOKS AND REVENUE) BY MONTH & CUMULATIVE						
	MONTHLY			CUMULATIVE		
	ACTUAL		TARGET	ACTUAL		TARGET
MONTH	#BOOKS	REVENUE	REVENUE	#BOOKS	REVENUE	REVENUE
Nov	9	90	100	9	90	100
Dec	10	100	100	19	190	200
Jan	6	60	100	25	250	300
Feb	7	70	100	32	320	400
Mar	8	80	100	40	400	500
Apr	5	50	100	45	450	600
May	13	130	100	58	580	700
June	11	110	100	69	690	800
July	15	150	100	84	840	900
Aug	12	120	100	96	960	1000

Gardens cookbook sales during the winter and spring months and may take action in the form of promotions if sales are below the target. The project manager may present this information in text format, or he might report number of books sold and revenue per month and cumulatively in table format as shown in Table 8.2. He might also compare the sales to target sales for this year and to sales per month for the prior year. While the project manager may find display of this information in text format easy to prepare and present, stakeholders may find it more difficult to assimilate the data in this format.

The table format provides a picture of the monthly and cumulative actual and target data for book sales and revenue. This is more concise than a narrative of this information, but still requires study and analysis for stakeholders to determine the success of the Community Gardens cookbook sales.

The project manager might create charts of monthly and cumulative cookbooks sales, including number of books sold and revenue compared to target data (Figures 8.1 and 8.2).

Stakeholders easily determine the status of Community Gardens cookbook sales, and the project manager uses this information to determine when to initiate promotions to increase sales and bring the actual cookbook sales closer to the target. For example, cookbook sales were trending below plan in the winter months. The project

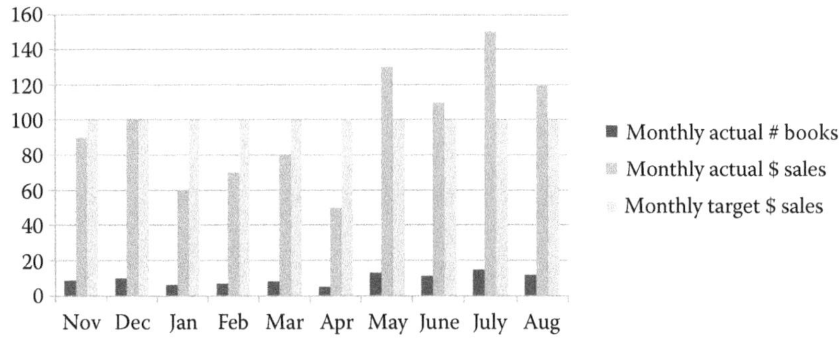

Figure 8.1 Community Gardens cookbook monthly data.

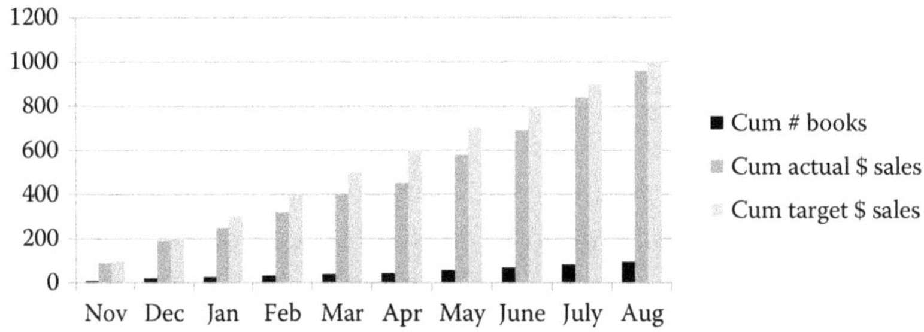

Figure 8.2 Community Gardens cookbook cumulative data.

manager, looking at the impact of low sales on cumulative data, might initiate a sales promotion in April, when sales had been trending below plan for several months.

The challenge is to present information in a format where it is easy for stakeholders to understand the impact of the data on the project objectives in the form of scope, schedule, budget, quality, and other project goals. This may require more effort from the project manager, but it enhances stakeholder understanding and engagement.

8.3.3 Meetings

The project manager incents shared ownership of the project using communication and decision-making tactics. In a project environment, the project manager addresses information from the perspective of the stakeholders. Project success is far more challenging when stakeholders view the project and project success as the responsibility of the project manager and project team; project success is easier when stakeholders view the project and project success as a joint effort and refer to *our* project rather than viewing the project as belonging to the project manager and team.

The project manager involves the stakeholders to generate this sense of shared ownership. The project manager engages the stakeholders by getting out of his own space; he talks to stakeholders in their language and meets with them on their turf. While the default language is generally English, there are variants based on areas of expertise. Gardeners may talk in terms of yield per plant and soil pH factors. The soup kitchen folks are more concerned with pounds of produce and nutritional value per pound. The 4H contest participants may talk in terms of color and size of the produce. The project manager and project team talk in terms of scope, schedule, budget, and quality. When working with different stakeholder groups, the project manager uses the language of the stakeholder audience to ensure communication and engage the stakeholder.

As part of project planning, the project manager schedules periodic meetings for updates, status checks, and forecasts of future project performance. Best practices for meetings require that he have an agenda for each meeting. An agenda is simply a list of topics to be discussed at the meeting. Figure R.1 contains a sample agenda for the Community Gardens Kickoff Meeting. If the project manager expects to make decisions at the meeting, he must ensure that attendees have background information necessary to make informed decisions prior to the meeting. Stakeholders need this information early enough so they can review, analyze, and make a preliminary informed decision prior to the meeting.

As the leader of the meeting, the project manager needs to make sure that the team stays with the agenda, that all participants have an opportunity to contribute to the discussion, and that conflicts are managed.

The project manager documents decisions made, additions, and updates to action items, and any other meeting highlights, including the date/time/location for the next meeting.

The project manager may also meet separately with one or two team members or stakeholders to share sensitive information or manage issues not appropriate to a wider audience.

8.3.4 Managing by Walking Around

The project manager engages stakeholders in different settings. MBWA, Managing By Walking Around, is a technique that captures this informal, but very effective, communication and stakeholder engagement tactic.

At times, the project manager schedules a meeting for a specific date and time in his office, home, or other venue. Other times, he may wander into the space of the stakeholders, such as the Community Gardens, the soup kitchen, or farmers market, and engage the stakeholders in an impromptu conversation on their turf. The project manager might select a neutral venue, such as a local coffee shop. Structured formal meetings may cover the topics on the agenda but may impede the sharing of problems or concerns. Conversely, informal conversations provide an opportunity for stakeholders to share concerns before they become problems and get in the way of project progress.

8.3.5 Sticky Notes

The project manager has many options for capturing information. He writes it down in a list format using pen and paper. He captures it in any number of automated ways—on a smart phone, tablet, laptop, or personal computer, each of which has many applications for capturing and storing this information. These approaches have different benefits and limitations. Handwritten materials are easy to create but are difficult to maintain, distribute, and keep current with changes. Content is easy to change and maintain with automated tools. However, there is only one person managing the information, and the audience tends to check out and have little sense of ownership of the information.

On projects, one of the many challenges is not only to use tactics that yield excellent results but also to use approaches that generate shared ownership of the results. If the project manager facilitates discussions and updates documents with little involvement of stakeholders, others in the room tend to have limited ownership. If the stakeholders think that someone else owns the project, they have little incentive to make the project successful. A better approach is to involve the stakeholders. If the project manager uses techniques that generate shared ownership of the results, the stakeholders are motivated to ensure that the project is successful.

The use of sticky notes is one tactic to incent shared ownership. If the team is creating a list of risks, the project manager can create the lists. Or, he might ask the team for ideas and add these to a list that he creates. A better approach might be to distribute sticky notes or index cards to the team and ask them to write an idea on each sticky note. The project manager then aggregates the notes across all the team members, organizes them into categories, and eliminates duplicates. Everyone sees the value of their contribution to the team work effort, and this shared ownership creates a powerful incentive to make the project successful. Project managers use the sticky note approach in the creation of key project documents including WBS, schedule network diagram, and list of risks.

8.3.6 Brainstorming

Once the project manager has gathered stakeholders at a meeting, he still has the challenge of getting the meeting attendees to actively participate in the discussions. Some team members are reluctant to share ideas, and some stakeholders don't like speaking in a public setting. Other times, dysfunctional team dynamics get in the way of proactive, positive sharing of ideas.

Brainstorming is the nonjudgmental sharing of ideas in a group setting. As stakeholders share ideas, the meeting facilitator captures these ideas on a flip chart or white board using the wording of the meeting participant. Other meeting participants may build on the idea, but there is no judgment or evaluation, good or bad, of the idea. People contribute their ideas verbally or write them on index cards or sticky notes, and the meeting organizers collect them into a list.

The brainstorming technique involves all meeting participants. It supports shared ownership of the project. The result is information used in the project plan.

The project manager is skilled in these different tactics and selects the most appropriate strategy.

8.4 Cross-Project Processes

There are processes that are necessary to successful project planning and execution but are bigger than any single project. These are similar to the standards discussed earlier; risk categories and probability and impact scales are used across all projects.

As listed in Table 8.1, there are several processes necessary for all projects to manage and control the challenges of team efforts. Change is frightening to many people. The status quo is familiar and comfortable while change is unknown and the consequences are uncertain. As a result, individuals and teams tend to resist change. Since projects are all about change, the project manager needs a process to introduce and manage change for the project stakeholders.

Decisions are necessary in the project environment. The project manager considers who should be involved in a particular decision and selects the decision-making style most appropriate to the project culture and the decision at hand.

Conflict is also a fact of life in project environments. The project manager is aware of the major sources of conflict on a project and is skillful at adapting conflict management techniques to the project environment.

Lastly, projects progress through a predictable sequence of phases. The project manager understands the activities necessary for project success in the various phases and actions to manage risks and potential problems.

These cross-project processes are as follows.

8.4.1 Change Management

Change management is a key project component and a necessary prerequisite to good project management. Trying to manage a project that lacks a change control process is similar to trying to execute a project that has no project plan. Stakeholders suggest changes to project scope without consideration of the impact on schedule, budget, quality, or other project components.

When there are no apparent consequences to making changes, stakeholders perceive that these changes are free. Stakeholders enjoy getting something for nothing; if there are no consequences to making changes, the project environment actually encourages stakeholders to ask for even more changes. If the project manager doesn't evaluate proposed changes for impact on scope, schedule, budget, and quality, he gives the impression that proposed changes are free; stakeholders are incented to ask for still more changes.

All projects need a change control process; it should be defined outside of any single project and used on all projects. If there is no change control process, the project manager develops one and makes it available for use on other projects.

The change management process includes a change request form, a process for evaluating proposed changes, a decision-making group authorized to make go–no-go decisions on proposed changes, a process for communicating the result of the change request, and if approved, a process to incorporate the change into the project plan. This includes updates to scope, schedule, budget, quality, and risks.

A change request form includes a description of the requested change, the requestor, date of request, and the date on which a decision is required. The evaluation of the requested change includes the benefits, limitations, impacts, and recommendations to approve or deny the request. This information may be a written request (in an email or letter), or a completed change request form. The project manager also identifies who will approve or decline the request. This might be a single stakeholder such as the sponsor or a group of stakeholders, often called change control board (CCB).

The process for reviewing and approving or declining the change request might be a meeting, either face-to-face, or virtual by phone, where the CCB balances the benefits of the proposed change against the costs and risks imposed by this potential change. The process specifies the maximum amount of time the CCB has to make a decision. The project manager or CCB notifies affected stakeholders of the decision in writing and updates appropriate sections of the project plan to implement the approved changes.

As a general practice, the project manager should *not* approve or decline the change request. It is better that the project manager is impartial and not potentially viewed by stakeholders as a gatekeeper.

The project manager evaluates the request, identifies several candidate approaches (usually three alternatives) to solve the problem addressed by the proposed change, identifies the pros and cons of the different approaches, and forwards this information to the CCB for consideration.

Until the project plan is finalized and formally approved, it is a work in progress and not typically subject to change control; changes are made as necessary during the iterative planning process. Once the project plan is finalized and formally approved, changes are subject to the change control process.

Figure 14.1 illustrates a change request for the deck project, and Figure R.2 illustrates a change request for the Community Gardens project.

8.4.1.1 Checkpoint Consider a change management process for your project. Make sure that you have specified an acceptable variance for key project objectives in the quality plan. For example, project schedule variances of plus or minus 1 week might be fine, but variances in excess of 1 week might require a change to the schedule. Or you might allow a 2-week variance early in the project but only allow a 4-day variance closer to project completion. Identify similar variances for your project budget and other significant project objectives. Identify the stakeholders who evaluate requested changes, the amount of time to make the decision, and the process for communicating the change to the affected stakeholders. Document this and include it in the project plan.

8.4.2 Decision Making

Decision making is an ongoing process in projects—stakeholders constantly weigh alternatives and consider the benefits and limitations of one course of action or another. Projects are people working in teams, and team decision making is challenged by different and often conflicting points of view.

When confronted with alternate courses of action, the project team needs to respond quickly. The project manager assesses alternatives, he makes a decision, and the project team implements the decision in a timely manner. If he delays making these decisions or fails to make a decision, partially implementing different alternatives, he causes project delays and often unnecessary costs.

Best project practices suggest that stakeholders define and agree on the decision-making process and tactics before the project starts. If the project manager has them processes from prior projects, he can simply reuse them, modifying the processes as necessary. If he doesn't have these processes from prior projects, he creates decision-making processes. The project manager then has them available for the current project and for reuse on future projects. In either case, the project manager needs to review the processes with project stakeholders and obtain their agreement. He wants all stakeholders to view these practices as shared project practices and not belonging solely to the project manager.

The rigor with which the project manager applies these processes depends on the importance of the decision; not all decisions require attention from the entire team, and there may be little risk to making less significant decisions without a formal process.

A process for decision making might include the following steps:

- Determine if the decision can be made by an individual (such as the sponsor or subject matter expert) or whether it requires review by a group of key stakeholders.
- If this is a team or stakeholder group decision, determine whether the group meets face-to-face or via a virtual meeting, such as a telephone call. Face-to-face tends to be the most effective communication technique, but it is sometimes challenging to get the stakeholders together. Telephone calls are easier to coordinate than face-to-face meetings, but it can be difficult to get everyone to participate. Instant messages are timely, but the team has no audit trail of the discussion. E-mail has the benefit of being asynchronous (people read and reply at different times) but strings of e-mail messages often trend off topic, consume a lot of people's time, and often take days or weeks to agree on a decision.
- Determine the alternatives and the benefits and limitations of each option.
- Make the decision and update any plans (schedule, budget, or action items) as appropriate.

Project managers rely on four basic decision-making styles.

The *command style* of decision making is characterized by one person taking charge and unilaterally making the decision. This may or may not include consultation with other team members or stakeholders. This style is appropriate when time is of the essence; there is no time to delay or consult with others. The risk with this style is that others may take offense that they were not consulted, they may not agree with the decision made, and they may not help to implement the decision. In fact, they may resist the actions necessary to implement the decision taken.

In the *consultative style* of decision making, the decision maker discusses the situation along with the alternatives (and the benefits and limitations of the alternatives) with key team members and stakeholders. The decision maker collects the necessary information, and based on the advice of the others, the decision maker makes a decision. This is different from the command style above in which time constraints preclude consultation with other team members while making the decision. The consultative style involves team members and key stakeholders so they have some ownership of the decision and the process. However, those who recommended a decision other than that selected may be offended and disengage from the decision implementation process. This style tends to work best when there is trust among stakeholders; everyone is free to contribute based on their expertise and trusts that the decision maker will do the right thing with the shared information.

The *consensus style* of decision making includes stakeholder involvement (as with the consultative style above) but instead of a single person making the decision, the entire team (or designated decision-making group) must agree on a course of action. This tends to create team ownership of the decision and the course of action selected. This style tends to work best when the quality of the decision is important. The decision maker not only needs all appropriate information shared but he also wants the collective evaluation and agreement of the stakeholders in the final course of action.

The *coin flip (random)* decision-making process is just that—random. This works when there are few or no criteria for more structured decision making, and the stakeholders are willing to accept whatever decision is made.

The project manager may choose other styles; he should select styles that will work best for the team and the project environment.

8.4.2.1 Checkpoint Think about your project and your project stakeholders. Based on the relative significance of the project, the project climate, and the nature of the stakeholders, select a decision-making style. Make sure that the stakeholders support the decision-making style selected and sign off on this important project process. Document the agreement in the project plan along with other standards and documents.

8.4.3 Conflict Management

Conflict is a fact of life—whenever people work together, they have different viewpoints and there is conflict. Conflict is neither good nor bad—it simply is. Often conflict about a course of action leads to heated discussion that causes an "ah-ha" moment for everyone in the discussion; this may result in a better course of action or decision than originally proposed by any of the people in the conflict.

Many people are uncomfortable with conflict and they try to minimize it or pretend that it doesn't exist. Conflict must be dealt with openly and sooner, rather than later. If conflict is left alone, it generally escalates and often involves more stakeholders, as each side to the conflict recruits support from other stakeholders. Sometimes, the conflict seems to go quiet with no evidence of the conflict at all. Generally in this situation, the conflict is building quietly and will eventually escalate and surface. It is easier to deal with the conflict early and before it escalates in tone and involves more stakeholders.

It's best if the stakeholders involved in the conflict work directly with one another. The project manager keeps conflict out of meetings, since an audience to a conflict sometimes encourages the stakeholders involved in the conflict to escalate it in front of an audience. If the stakeholders involved in the conflict cannot resolve it among themselves, the project manager may need to get involved. This generally takes the form of a meeting with the stakeholders involved in the conflict with the project manager mediating the discussion.

The project manager anticipates and manages conflicts on the project. There are many sources of conflict on projects; the proactive project manager promotes shared ownership of plans and decisions in areas subject to conflict and contention.

8.4.3.1 Checkpoint Think about your project and your project stakeholders and identify a process for managing conflict. Discuss this with the key stakeholders and document it in the project plan. Since each conflict is different, the conflict management approach may be general and adapted to each individual conflict circumstance.

8.4.4 Project Life Cycle: Planning versus Executing

Projects go through a series of phases, called a project life cycle. For our personal projects, this book explicitly focuses on the planning and execution phases of the project life cycle.

A project life cycle typically includes phases of approval and start up, planning, executing, tracking and fixing, and closing. In the *approval and start up phase*, the project is approved and the project manager and project team begin to plan the project; this phase authorizes the expenditure of additional effort to plan the project.

In the *planning phase,* the project manager and project team figure out what needs to be done to complete the project and accomplish the deliverables. The project manager determines overall project feasibility before he starts the work of the project. An example might be building a house. It makes little sense to hire someone to dig and pour the foundation if there is no blueprint for the overall structure. Without a plan, the project team might spend money digging and pouring the foundation and then realize that they don't have the time, money, or resources to complete the rest of the structure. It's better to plan the entire structure first and then contract for a foundation that is small enough so that the project has sufficient time and resources to complete the rest of the structure.

In the *executing phase*, the project team does the work of the project; they implement the plans previously documented. Typically, this phase takes the most time and costs the most money. When the project manager does a good job of planning, the execution is relatively easy because the difficult choices and decisions have already been made. Unfortunately, if the plans are not complete or clear, then the project team experiences a number of false starts when they execute and they often have to repeat work that is misdirected, incomplete, or simply wrong.

While the team is executing (doing the work of the project), the project manager compares the actual execution to the plan on an ongoing basis (*tracking and fixing*). When the project is on-plan within specified parameters (for example, plus or minus 1 week of schedule, plus or minus $x of budget), then the project manager notes the positive project progress and continues project execution. However, if the actual-to-plan comparison exceeds acceptable variances, the project manager takes action. He either changes the implementation tactics to bring the project back in line with plan, or, going through a change control process, the project manager revisits the plan and makes necessary changes. These necessary changes generally take the form of reduced scope, more time, and/or more money or resources.

Once the project team has completed the work of the project, the project manager formally *closes* the project. All key stakeholders sign off on the work of the project and accept the product of the project. The project manager also conducts a lessons-learned meeting, in which he reviews the project with stakeholders. He identifies what worked, what didn't work, and what he would do differently on future projects. Figure 14.2 illustrates lessons learned for the deck project, and Figure R.3 illustrates lessons learned for the Community Gardens project.

There is an additional informal phase that occurs after project close. The quality plan contains a column for actual performance with respect to the listed project and product metrics. Some of the metrics are assessed at project completion; other metrics cannot be assessed until the product of the project has been in use for some time. The project manager revisits these quality metrics at the appropriate time post project and finalizes the quality plan with this information.

These phases enable an orderly progression of project work, while minimizing false starts, wrong directions, and unapproved efforts. This level of process is necessary where there is significant risk with respect to scope/quality, schedule, and budget; the project manager may abbreviate the rigor with which he implements the life cycle phases in smaller, less risky personal projects.

At the end of each phase, the project manager meets with the appropriate stakeholders to review the work of the phase. If all the project and product work is complete and satisfactory, the team approves the completion of the current phase and progression to the next phase.

These phase completions are significant project milestones; the project manager includes them on the project schedule.

8.4.4.1 Checkpoint Think about your project and determine the project phases necessary to project success. Document this project life cycle and obtain the approval of key stakeholders. Revisit key project documents, such as the WBS, schedule, and risk plan, and ensure that they reflect the phase completions and other impacts of the selected project life cycle.

8.5 Project Toolkit Summary

The project manager relies on the project toolkit to facilitate project planning and project execution. It provides proven document templates, standards, tactics, and processes that enable reliable and predictable project performance.

There are many additional documents, standards, tactics, and processes that a project manager may find useful. Pareto analysis and cost–benefit analysis may be used for prioritization of projects and project features. Decision trees are useful for exploring and weighting alternatives. Rolling wave planning is useful for long-range planning. SWOT (strength, weakness, opportunity, threats) analysis is useful for classifying risks, and expert judgment is excellent for validating team decisions and courses of action. The reader is encouraged to explore additional toolkit items.

As project stakeholders become familiar with the toolkit, they share ownership of the process and the project plan with the project manager and the project team. Shared ownership of the project plan promotes active engagement and enhances project success.

The chapters that follow focus on the project context. The project core (scope, schedule, budget, quality, and risk) provides the key content of the project; the project toolkit provides the tools to enable project planning and execution. These occur in a context that influences the stakeholders' success. It's time to look at the context.

9
Project Context

Everything takes place within a context. The very same challenge may be more or less difficult based on this context. For example

- A plant that requires a lot of moisture doesn't do well in the desert; a cactus doesn't thrive in a rain forest.
- Freshwater fish die in saltwater; saltwater fish die in fresh water.
- Athletes train in high altitude, low oxygen environments, to increase lung capacity and increase the probability of doing well in events at sea level.

Similarly, projects take place within a context. The very same project may be more or less difficult based on its context. The project manager proactively considers the project context and determines how this contributes to, or gets in the way of, project success. He then leverages and mitigates these environmental factors.

The core project includes scope, schedule, budget, quality, and risk. The project context surrounds the project core and includes project stakeholders, project communications, and teaming. The project needs resources, including people and equipment. Depending on the size of the project, the project manager may deal with a large number of people, including customers, sponsors, and team members who execute the work of the project. All these people have a stake in the project; they are project *stake*holders. They may or may not have worked together in the past. They may work together well, or there may be a lot of tension among the team members. The project manager creates a context in which the team members work effectively together. He ensures that communications among team members are effective and productive. He has tactics to address miscommunication and conflict among team members and proactively takes steps to build a cohesive team.

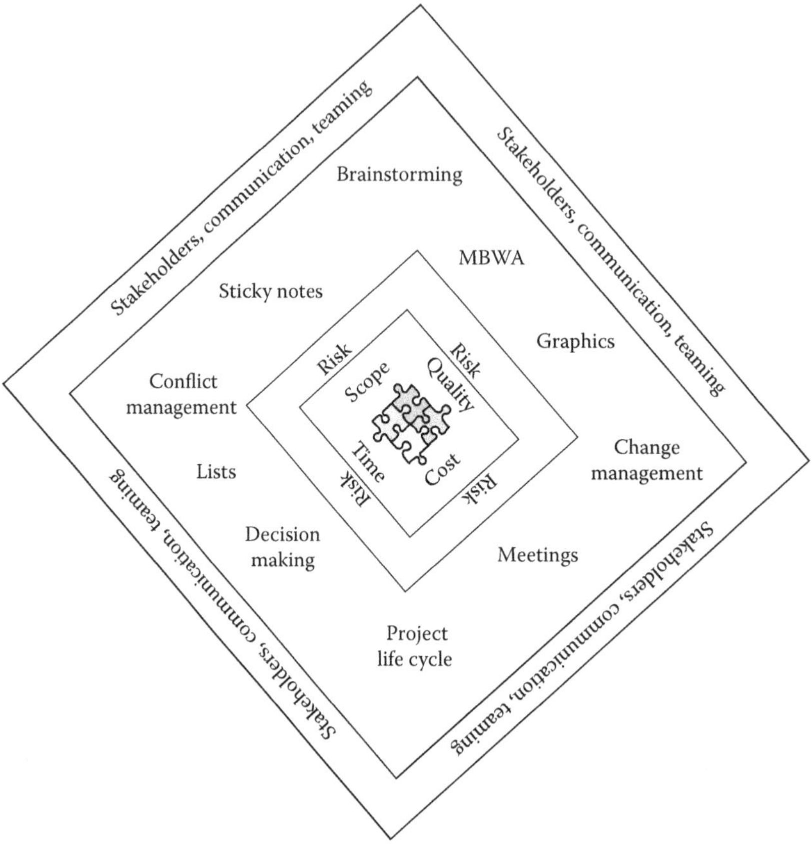

Figure 9.1 Project environment.

Stakeholder management, the range of communication issues, and the necessary teaming vary based on the size and importance of the project, the number of stakeholders, and team maturity. The project manager considers the specific project and the project complexity and addresses these components at a level commensurate with the significance of the project.

Figure 9.1 illustrates the stakeholders, communication, and teams as the context in which the project manager plans and executes the project core. Between the project core and the context of stakeholders, communication, and teaming are project toolkit components discussed in the prior chapter.

The project context including stakeholders, communication, and teaming are discussed in the chapters that follow.

10
Project Stakeholders

Stakeholders are a key component of the project context. As with all components of the project context, project success or failure depends on the project manager's and project team's ability to identify, plan, and manage this larger environment in which the project exists.

Stakeholders have a *stake* in the project; they are critical to project success and include the resources that

- Do the work of the project
- Identify project requirements
- Consume the product of the project

A stakeholder is anyone positively or negatively affected by the project. This is a broad definition and includes many people.

In the deck project, stakeholders include:

- The project manager.
- The property owner, who is the sponsor as he provides the money for the deck, and is a customer as the property owner.
- The renters are customers as they will use the product of the project, the deck. They may also be considered sponsors if they assist in the construction of the deck.
- Abutters' property is adjacent to the property on which the deck is built. They are affected if the deck is too close to the property line, if the construction is noisy or takes place late at night or early in the morning, and if the renters have loud parties that extend late into the evening.
- Town permit department. The town is the recipient of the permit application, conducts the property inspection, and grants or denies the permit to the property owner. The town also receives the application fee.
- Vendors of building supplies. Vendors provide building materials as specified by the project manager.

The project manager and the project team pay attention to all stakeholders. If they only focus on one category of stakeholder, they risk losing the support of the other stakeholders who can make the project a success or limit its impact. The goal with respect to stakeholders is to keep positive stakeholders happy and to change negative stakeholders to either positive or neutral. To accomplish this, the project manager and the project team analyze the stakeholders and develop strategies to align the interests of the stakeholders with the goals of the project and the product of the project. Since stakeholders may change over time, the project manager continues to analyze and manage the project stakeholders over the life of the project.

10.1 What's in It for Me?

Self-interest motivates everyone. People do things to feel good (think massage), to generate something they value (money, positive recognition, positive self-esteem), or to avoid pain and negative consequences. Avoiding pain includes not eating foods to which we are allergic. Negative motivations might include studying to avoid poor grades or working unpaid overtime to avoid losing a job. Stakeholders behave in ways that are consistent with their self-interest.

Since the project manager influences stakeholder behavior (keep them positively inclined toward the project or change their project bias from neutral or negative to positive), he needs to understand their self-interest with respect to the project. This is WIIFM—what's in it for me. If the project is successful, what does the stakeholder gain? If a stakeholder is negatively biased toward the project, the project manager wants to understand what the stakeholder loses if the project is successful. Understanding the stakeholder WIIFM, the project manager anticipates and influences stakeholder actions and attitudes toward the project.

If the stakeholder stands to gain if the project is successful, it may be sufficient to highlight how project success benefits the stakeholder to gain stakeholder support. If project success threatens the stakeholder, the project manager understands that this perceived threat is the reason why he wants the project to fail. The project manager identifies steps to minimize the negative impact or find a silver lining.

Stakeholders also influence each other. Renters share their enthusiasm for the deck with neighbors and get them excited about possibly building their own deck. The project manager reassures neighbors that he limits construction to regular working hours so not to disturb neighbors early in the morning or later in the evening. He also reassures the neighbors that parties on the completed deck will be subdued, and there is substantial distance between the deck and the property lines. The property owner, also the project sponsor, influences the town building department with respect to the excellence of the deck construction and adherence to building safety guidelines. He might also work with vendors to obtain discounts on building supplies. In this small project, the project manager also does most of the work; therefore, he is also the project team. He might enlist the help of the other renters (his family or roommates) to do the work of the project and the product, expanding the size of the project team.

10.2 Identify Project Stakeholders

Since project stakeholders influence the project, it is important to understand their bias and positively influence them. A pre-requisite to influencing stakeholders is to know who they are. The project manager and project team identify the project stakeholders and create a list. They gather this information from the stakeholders listed in the project documents including the charter, scope statement, resources assigned to tasks, and risks. The project team might also brainstorm to identify additional stakeholders and ask stakeholders to suggest others that might be involved in the project.

For each stakeholder, the team collects and documents key information and organizes this information in a stakeholder matrix, which contains the following information:

- Name
- The stakeholder's perceived power over the project and interest in the project from the Power-Interest Grid on Figure 10.1 (discussed in Section 10.3)
- The stakeholder's priority based on the combined power–interest score

- What the stakeholder needs from the project—each stakeholder determines his needs from the project
- What the project needs from the stakeholder—the project manager determines what the project needs from each stakeholder

In most projects, the project manager has more stakeholders than he has time to manage so he prioritizes the stakeholders. The project manager spends more time with those stakeholders that can help or hurt the project the most. The priority determines the most influential or powerful stakeholders; it is a blend of their power and interest.

Table 10.1 illustrates the completed stakeholder matrix for the deck project. The project manager completes the first column with the list of stakeholders. Later, working with the team and the power interest grid discussed below, he completes columns 2 and 3 with information about stakeholder power, interest, and priority. He then organizes the list of stakeholders by priority and creates a relative ranking with stakeholders with higher priority scores ranked higher than those with lower priority scores. For the top-ranked stakeholders, he completes columns 4 and 5 with information about what the stakeholder and project need from each other. If necessary, the project manager meets with the various stakeholders to understand their WIIFM from the project to complete these columns.

Appendix M contains the stakeholder matrix for the Community Gardens project.

10.2.1 Checkpoint

Create a list of stakeholders for your project, completing *only* the first column of the stakeholder matrix.

10.3 Organizing the Stakeholders

The project team creates a list of stakeholders from the project documents and identifies additional stakeholders through brainstorming and discussions with other project stakeholders. The team copies these stakeholders to sticky notes, with one stakeholder per sticky note, and places the sticky notes on the Power Interest Grid in the appropriate

Table 10.1 Stakeholder Matrix for Deck Project

STAKEHOLDER NAME	POWER (P) INTEREST (I)	PRIORITY	WHAT STAKEHOLDER WANTS/NEEDS FROM THE PROJECT	WHAT THE PROJECT NEEDS FROM THE STAKEHOLDER
Project manager	High P High I	1	Complete project on scope, schedule, budget	Direction and communication
Property owner	High P High I	1	Deck completed per scope, quality, and budget	Funding, requirements, and timely approvals
Renters	Low P High I	3	Deck completed on schedule; deck is safe	Support, positively influence abutters
Abutters	Low P High I	3	No noise, intrusions during construction; no noisy parties once complete	Support; no objections to permitting
Town permit org	High P Low I	2	Proper paper work submitted	Inspection and permit
Vendors of bldg supplies	Low P Low I	4	Profitable business	Quality and inexpensive building supplies

quadrant. The sticky notes make it easy to move a stakeholder from one quadrant to another as discussions cause the team to rethink which quadrant is most appropriate for a stakeholder.

Stakeholders fall into one of the grid sections as follows:

Priority 1 = *High Priority* based on high power and high interest
Priority 2 = *Medium High Priority* based on high power and low interest

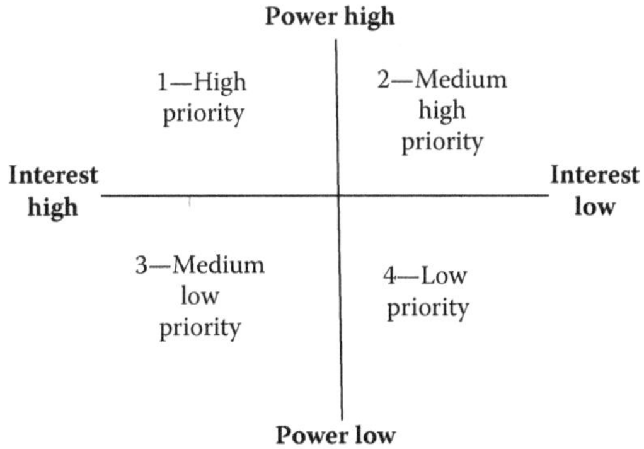

Figure 10.1 Power interest grid.

Priority 3 = *Medium Low Priority* based on low power and high interest
Priority 4 = *Low Priority* based on Low Power and Low Interest

The project manager uses the priority to determine which stakeholders are likely most influential and require more attention from the team. The project manager may consider all stakeholders with a priority of 1 or 2 significant. Or he might take the top 40% of the stakeholders, regardless of their priority score. The project manager completes columns 4 (what the stakeholder wants/needs from the project) and 5 (what the project needs from the stakeholder) for these most significant stakeholders.

Some stakeholders may view the priority rankings as political, with a lower priority perceived as negative. They may feel that they are not as important as stakeholders with a higher priority and become alienated from the project. To mitigate this, the project manager involves stakeholders in the prioritization process and discusses the final prioritization with all affected stakeholders. He explains that a low ranking is not negative; it is simply an indication of this project relative to other projects and other stakeholders. The discussion focuses on the stakeholder's power with respect to the project and the stakeholder's interest in the project; the priority is derived from these variables, it is not assigned directly. The project manager assigns these priorities with sensitivity and political awareness.

For the deck project, both the project manager and the property owner (sponsor) are high power and high interest. The property owner controls the funding and is concerned about the impact of the deck on his property and rental value. The renters and abutters are very interested in the impact of the deck on their ability to enjoy their outdoor activities; both stakeholder groups have limited power as they do not contribute money or resources. The renters have some power, in that they may support the project manager in the deck construction. Abutters may have relationships with people in the town building department and may try to influence the building department to deny the permit if they are concerned about noise, proximity of the deck, or other intrusions. The town permitting organization has high power, in that they can deny the permit; they have little interest in this particular permit request. Vendors are interested in obtaining the business, but it is not likely high dollar value so their interest is limited. They have no power over this project.

The project team posts project information on the project website (or other publicly available site like Facebook), and stakeholders with low priority access project information themselves. Stakeholders with high interest are likely to seek out project information and proactively advise the project manager if there are issues or opportunities. Although the project manager assigns a relatively high priority to stakeholders with high power and high interest, he is confident that these stakeholders will let him know if there are concerns. Stakeholders with high power and low interest are a challenge. High power enables them to significantly impact the project. Low interest suggests that they may not pay attention to the project. The risk is that they will not react until the project is too far off course for the project manager to correct problems and bring the project back on track. To mitigate this risk, the project manager prioritizes these stakeholders and puts a communication plan in place where he proactively reaches out to these stakeholders. He meets regularly with these stakeholders so he can discover early if things are not going according to their expectations. Table 10.1 contains these power interest scores for the deck project stakeholders, and Table M.1 contains this information for the Community Gardens project stakeholders.

10.3.1 Checkpoint

Work with your team to assign stakeholders to one of the 4 quadrants in the power interest grid and to assign a priority based on the quadrant. Add the power interest scores and stakeholder priority to the stakeholder matrix. If your project has many stakeholders, rank order the stakeholders based on either their priority (for example all priority 1's or all priority 1's and 2's) or the top X% of the stakeholders. Complete column 4 (What stakeholder wants/needs from the project) and column 5 (What the project needs from the stakeholder) for the top-ranked stakeholders.

10.4 Stakeholder Engagement

To maximize the probability of project success, the project manager prefers that all stakeholders are positively inclined toward the project.

While this may not be realistic, the project manager tries to have many stakeholders positively biased toward project success and few stakeholders neutral or negatively biased toward project success. The project manager determines the stakeholder's bias today, the preferred future bias, and identifies a strategy to move key stakeholders from today's current state to the preferred future state.

The following are possible stakeholder engagement levels:

- *Unaware:* The stakeholder is unaware of the project and not concerned with the project impact. This is neither a positive nor negative bias toward the project.
- *Negative:* The stakeholder is aware of project and perceives the potential impact to be negative. The stakeholder is likely resistant to the changes introduced by the project. This is considered to be a negative bias.
- *Neutral:* Unlike the unaware stakeholder, these stakeholders are aware of the project but are neither positively nor negatively biased toward the project.
- *Positive:* These stakeholders support the project and the changes expected as a result of the project. This is considered a positive bias.
- *Champion:* These stakeholders actively advocate on behalf of the project. They fully support the project and the changes expected as a result of the project. They actively promote the project to ensure that the project is a success. This is considered a positive bias.

Working with the team, the project manager determines today's bias (T) and the preferred (P) future levels of stakeholder engagement for the deck project (Table 10.2).

Using this matrix, the project manager easily sees the current and preferred levels of stakeholder engagement. His challenge is to move the stakeholders from the columns on the left to the columns on the right, which are categories more supportive of the project. The project manager targets his effort toward those stakeholders that can help the most or hurt the most. This would be champions and negative stakeholders with high power, respectively.

The project manager selects the top-priority stakeholders and creates a strategy to move those stakeholders from the current level of

Table 10.2 Stakeholder Engagement Matrix for Deck Project

STAKEHOLDER NAME	POWER INTEREST	PRIORITY	UNAWARE	NEGATIVE	NEUTRAL	POSITIVE	CHAMPION
Project manager	High P High I	1					T P
Property owner	High P High I	1				T P	
Renters	Low P High I	3				T P	
Abutters	Low P High I	3		T	P		
Town permit org	High P Low I	2	T P				
Vendors	Low P Med I	3	T P				

Notes: P = Preferred Level of Engagement; T = Today's Level of Engagement.

engagement to the preferred future level. It is more difficult to move a stakeholder several columns to the right, than to move him 1 or 2 columns to the right. These strategies will take effort, and the project manager revisits the work breakdown structure and adds the necessary tasks to the task list and allocates the necessary time in the project schedule.

For the deck project, the key stakeholders are reasonably aligned with respect to current and desired level of engagement. The abutters are currently negatively inclined toward the project, and the project manager prefers that they be neutral or positive toward the project. As noted above, the project manager assures the abutters of the limitations on working hours and tenants promise to avoid noisy parties once the deck is complete.

Appendix N contains the Stakeholder Engagement Matrix for the Community Gardens.

10.4.1 Checkpoint

Complete the stakeholder engagement matrix for your project. If there are many stakeholders, complete the matrix for the most significant stakeholders based on priority scores from the power interest grid. For the less significant stakeholders, complete only the first three columns of the matrix. Keep them on the list in case their priority changes and as a reminder to periodically review their priority.

The stakeholder information is the basis for the communication plan in the chapter that follows.

11

Project Communication

Imagine trying to work on or manage a project where everyone is in a separate room with no way to communicate with the other team members and stakeholders. Maybe everyone is working from home with Internet connections that are slow and noisy with interference. Imagine this same project where the only means of communicating with others is writing messages on index cards and passing them around. This would be time consuming, ripe for misunderstandings, and difficult for all stakeholders. This project environment is not much fun, and project progress is likely slow and error prone.

Timely and effective communication is challenging in all of these situations; it is also a critical success factor for project success. It is important that all stakeholders are aware of the information and its impact on project progress; it is less important whether the information is positive or negative.

A generally accepted rule of thumb is that 90% of a project manager's time is spent communicating. All project stakeholders spend significant time communicating, including the time:

- Spent in meetings
- Reading and writing e-mail
- Developing and reading project documents, including status reports, schedules, budgets, risk plans
- Spent in informal conversations building relationships and discussing technical issues

Consider the time required for communication if every stakeholder needed to communicate with every other stakeholder. The team would likely get nothing done but communicating with other team members. The formula to calculate the number of communication channels (NCC) is as follows:

$$\text{Number Communication Channels} = \frac{[N * (N - 1)]}{2}$$

where N = the number of people.

If a project manager has a team of 5 people, there are 10 communication channels. If he doubles the size of the team to 10, the NCC increases by a factor of 4.5 to 45 communication channels. Sometimes adding people to the team actually slows down the progress, rather than speeding it up; team members spend more time communicating, and less time is available for project work (Figure 11.1).

11.1 Communication Plan

Stakeholder communications are often informal with little planning concerning the content and presentation of the information to be shared. In the project context, the project manager and project team plan many of these interactions in advance. The project manager develops a communication plan as an extension to the stakeholder matrix.

For the top-ranked positive and negative stakeholders from the stakeholder matrix, the project manager captures the following information for the communication plan:

- Contact information including e-mail, phone, IM, twitter, Facebook, LinkedIn, and other appropriate contact information
- What information they need
- Time frame and frequency
- What the project expects from them, such as sign off, review and comment, or other action
- Who is responsible for generating the information

Once the project manager creates this communication plan, he shares it with the project stakeholders and obtains agreement from the stakeholders that their section of the communication plan will meet their information needs. The project manager reviews the communication plan periodically to ensure that it remains relevant as stakeholders change during the life of the project. He makes changes as necessary.

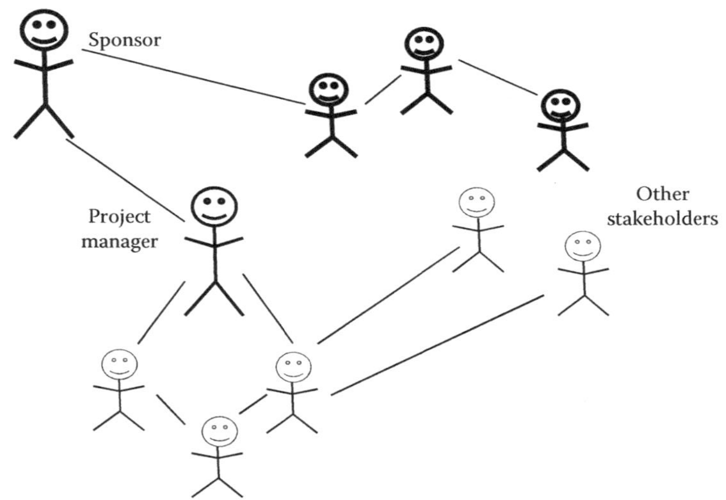

Sponsor's team communication channels
Team members in bold (includes PM)
NCC = [N * (N – 1)]/2; N = number of people
NCC = [5 * (5 – 1)]/2 = 10 communication channels

Project manager's team communication channels
Includes PM and nonbolded stakeholders
NCC = [N * (N – 1)]/2; N = number of people
NCC = [6 * (6 – 1)]/2 = 15 communication channels

Combined teams communication channels
Includes all stakeholders
NCC = [N * (N – 1)]/2; N = number of people
NCC = [10 * (10 – 1)]/2 = 45 communication channels

Figure 11.1 Stakeholder relationships.

11.2 What to Communicate

Since communication is so critical to project success, it is important that the project manager ensures that necessary information is captured and shared appropriately. The project manager must also balance the cost and time for communications against the value of the communications and the risk of not sharing information. The general rule is that the *amount* and *formality* of project communication should be appropriate to the importance and risk of the project. A low-budget project that will be completed in a relatively short amount of time with little risk requires less communication and formality than a project that risks more money and has a significant impact if it fails.

Amount of communication focuses on the frequency, such as weekly, monthly, or quarterly team status reports. For a relatively short 6-month project, the project manager may plan weekly status reports; for a multiyear project, the project manager may plan bi-weekly status meetings, changing to weekly or monthly as project conditions warrant. For a high priority, high risk project, the project manager may plan 10% of stakeholder time for communication. For a lower priority project, communication may consume 7% of stakeholder time.

While it is more costly and time consuming than informal verbal communication, there are a number of advantages to formal, written communication. It documents the stakeholder agreements, provides a basis for changes, and exposes misunderstandings among project stakeholders. The writing itself forces the author to think through and clarify his message and often exposes questions and information requiring clarification from project stakeholders. Written schedules help team members avoid overcommitment of time and to plan for additional project resources when necessary.

Communication is most effective when the project manager and project team speak the language of their audience. They use vocabulary and speaking styles appropriate to different audiences such as peers, managers, town officials, police, and vendors.

To complete the project communication plan, the project manager considers the information that stakeholders need for both project planning and execution. The responsibility assignment matrix (RAM) for project documents (see Chapter 12) contains much of the information required for planning. Rather than repeat this information, the project manager considers the RAM as an extension to the communication plan. The communication plan in Table 11.1 lists information necessary during project execution.

Appendix O contains the communication plan for the Community Gardens project.

11.2.1 Checkpoint

Working with the project team, complete the first two columns of the communication plan; the stakeholder matrix contains a list of stakeholders for column 1. Complete additional columns when the team has this information.

11.3 Push or Pull

Another consideration is how to get the information to the appropriate stakeholder. In the stakeholder chapter, the project manager recognized that high interest stakeholders will likely seek out project information, while low interest stakeholders may not pay attention to project information even when it is sent to them. As a general rule, the project manager pushes project information to high power, low interest stakeholders but may allow high interest stakeholders to pull project information from specified project information repositories.

Push communication is when the project manager pushes information to the intended recipient. E-mail is an example of push communication. Pull communication is when the project manager posts the information (e.g., on a web site or Facebook page) and lets the stakeholder access the information at his convenience. In the case study, the project manager might state in the communication plan that he will post updates for gardeners every Sunday evening by 8 p.m. Then the gardeners know to access the web site Sunday evenings or later and obtain the latest information.

Interactive communication is when all parties to the communication are present and exchange information in real time. Interactive communication is most effective for shared understanding; it provides an immediate opportunity for questions and to clarify potential

Table 11.1 Communication Plan for Deck Project

RECIPIENT OF INFORMATION	CONTACT INFORMATION	INFORMATION NEEDED	TIME FRAME AND FREQUENCY	ACTION = SIGN OFF (S), REVIEW (R), EXECUTE (E)	WHO CREATES THE INFORMATION
Project manager Priority = 1	PM@gmail.com 999-888-7766	Status reports scope, time, cost	Fridays by 5 p.m.—wkly	R	Project team (may be PM acting as team)
Property owner (sponsor) Priority = 1	PO@gmail.com 999-777-6655	Status reports scope/quality, cost	Mondays by 5 p.m.—wkly	S	PM sends information
Renters Priority = 3	R@gmail.com 999-666-5544	Alerts when delivery, bldg. intrusions	As needed	R	PM
Abutters Priority = 3	Multiple	Pre-, mid-, & post-project updates	Evenings	R	PM
Town permit org Priority = 2	TPO@gmail.com	Completed permit application	Before construction	S	PM
Vendors Priority = 3	Multiple	Details of bldg. materials	As needed	R	PM

misunderstandings. The challenge is that all parties to the communication must be available at the same time.

11.3.1 Checkpoint

As you complete the communication plan for your project (columns 3–6 on Table 11.1), decide how you will communicate the project information (formal/informal, written/verbal, frequency, push/pull/interactive). Identify the stakeholder responsible for generating the information and the stakeholder that will act on the information (actions might include sign off, review, execute, or other actions that you create for your project). Obtain stakeholder approval of the communication plan and save it in the project plan.

11.4 Marketing and Branding

Projects tend to do better and be more positively received, when stakeholders understand the complete project picture, rather than just one view of the project. The deck project is relatively small with few stakeholders. There is not a clear audience for a marketing and branding message.

The Community Gardens project has a number of constituencies, including abutters, soup kitchen and customers, farmers market and customers, 4H competitors, and the town manager. Consider the families that live near the Community Gardens and experience a significant increase in traffic due to the gardens. They may view the project negatively based on this limited view. They are more likely to view the Community Gardens positively if they have the added perspective of the soup kitchen and fund raising to help the less fortunate in the community.

The project manager cannot assume that people have the larger perspective of the project. He must proactively communicate the larger picture and the community WIIFM (what's in it for me). The project manager thinks about the message he wants to communicate (branding) and how he will get this message to the larger community (marketing). Maybe the message for the Community Gardens is "Eat Healthy, Support the Soup Kitchen, and Participate in the Farmers Market." Once the project team has the message (approved by key project stakeholders), the team needs a plan to market the project. This might include an article in the town paper, flyers posted around town, presentations at schools, senior centers, town hall meetings, and other relevant community gatherings. These should be included in the communication plan.

11.4.1 Checkpoint

Think about the marketing and branding for your project. Document this and update the communication plan to include this information.

11.5 Next Steps

The project manager next merges the various stakeholders into a dynamic and productive work force. This is discussed in the following chapter on stakeholder teaming.

12
Project Teaming

The chapter on stakeholders focuses on engaging the stakeholders in the project. The chapter on communication addresses the project and product information that will be generated and shared; it also displays tables for organizing the relationships among the stakeholders and the information. As the next step, the project manager organizes the stakeholders into a team.

Engaged stakeholders are a necessary, but not sufficient, prerequisite to team development. Stakeholders must be engaged with the project but they must also be engaged with other stakeholders.

While stakeholders may think the project and the product of the project are good ideas, this doesn't guarantee that they work well together to manage the project and develop the product of the project. Once the project manager has the stakeholders engaged, he assesses the extent to which they are a team and identifies techniques to get these various individuals working well together as a team.

Teams go through predictable stages of team development, and it is useful to understand these stages and the behaviors characteristic of the project manager and the team during these stages.

Using stakeholder and team information, the project manager develops project organization charts and responsibility assignment matrixes (RAMs). The organization chart identifies the relationships and roles of project stakeholders to each other; the RAM identifies the relationships of stakeholders to project deliverables.

The project manager includes these documents in the project plan and updates the project schedule and budget with time and money for necessary team-building activities.

12.1 Teaming

Teams generally do project work; projects are not an individual effort, they are a group effort. Team members bring different skills and viewpoints to the project which leads to better decision making. Properly managed, these diverse views and skill sets lead to more effective project processes and better products of the projects. Left unmanaged, these diverse skill sets and viewpoints may degrade into conflict and chaos. The challenges for the project manager are to create a team that works well together and to have proven techniques for dealing with conflict when it happens.

Bringing together a random group of people to work together doesn't make them a team. Sometimes the people on the team have worked together on prior projects and already established working relationships that may be effective or dysfunctional. Other projects are staffed by stakeholders that have never met and sometimes are geographically remote to each other. The project manager has the additional challenge of forming a functional team from a group of strangers and still greater challenges when creating a team from people that are geographically remote. This latter type of team is called a virtual team where the team members do not work in the same physical location.

The project manager determines if people on the team have worked together in the past and assesses the extent of positive or negative teaming among team members through observation and discussion.

Think about your experience observing a group of people working (or not working) together. It might be a business meeting, a book club, a family at dinner, or a ball team. The observer often discerns whether the team dynamics are positive or negative within a minute or two of observation and often without hearing any dialogue. How does he do this? He observes body language, where people sit relative to other team members, style of dress and personal hygiene, tone of voice, words and phrases used to communicate, eye contact, and a variety of other verbal and nonverbal signals. The observer then tests these observations with follow-up meetings and discussions and takes action to address any team dysfunctions.

Virtual teams pose a greater challenge. Most communication among virtual teams is remote, relying on phone, e-mails, online presentations and webinars, text messages, Skype-type messaging, and other communication techniques for stakeholders physically removed from each other. Sometimes, geographically centralized teams behave like virtual teams, in that they work from home or individual offices and choose to communicate with e-mails or other electronic techniques rather than face to face, eliminating the cues available with nonverbal communication.

Fifty-five percent of communication is non-verbal; in virtual teams, the available verbal techniques limit communication to 45%—a significant decrease over the communication available in colocated and geographically centralized teams. Teams that rely on electronic communication lack the cues available in face-to-face communication thereby risking more miscommunication.

The project manager determines the extent of teaming using verbal and nonverbal cues and manages the progression of the team through the process of team development discussed below.

12.2 Stages of Team Development

Teams go through a predictable process of development. Think about a recent class or team that you joined. Initially, everyone is polite, reserved, and looking to the teacher or coach for direction. After just a small time together, the team members become more comfortable with each other and begin to interact among themselves and rely less on the team leader.

Bruce Tuckman developed Tuckman's Stages of Team Development to describe the stages that people go through in the process of developing into a team. These stages are forming, storming, norming, performing, and adjourning. In the *forming* stage, team members get to know each other; they tend to be on their best "company" behavior and defer to the team leader. Once team members get past the polite behavior, they begin to jockey for position—what is their role, who reports to whom, and what are their responsibilities for project deliverables. The team members *storm* until they sort out roles, work responsibilities, and relationships. This stage is often characterized by conflict, and the team leader helps the team members work through the conflicts. Once the team sorts out these conflicts, it works together to establish rules of behavior and team norms (standards, guidelines). This is the *norming* stage. With roles and guidelines, the team self-directs, with the team leader in the background providing support, and the team *performs* at a high level, accomplishing the work of the project. The team leader is in the background, ready to step forward if the team needs direction. The team leader looks into the future, identifying potential roadblocks to the team's work and removes these roadblocks before they get in the way of the team's progress. Once the team completes the work of the project, the project manager closes out the project and the team *adjourns*, with team members moving on to new projects.

As the team goes through this development process, the team leader, acting as project manager, uses tactics to enhance and facilitate team building.

12.3 Team Building

In personal projects, teaming is challenging and perhaps more difficult than in a business environment. The project manager is often working with volunteers, so there is no monetary incentive, and the project is often competing for their time against family and paid work activities. If the project manager can make the project environment a fun place to be, then volunteers may look forward to the project work and spending time with the project team. The challenge is to make projects a fun and desirable place to spend time. The project manager uses nonproject work to create an environment where team members get to know each other and develop positive working relationships.

In the business environment, some team-building techniques and activities include:

- Nonproject activities including sports (bowling), lunch or dinner out of the office
- Celebration for project or phase completion
- Rewards and recognition for exceptional efforts

Activities that may be effective for the teams on personal projects include:

- Refreshments at meetings
- Public recognition for work well done or exceptional effort above and beyond expectations
- Recognition in a newsletter or local paper
- Continuing education credits toward maintaining professional certifications
- Credit in lieu of classroom work for degree programs

- Acknowledging altruistic motivations, sense of returning value to the community, or paying forward

Once the project manager identifies the team building activities for his project, he adds these to the list of tasks and adds time to the schedule and money to the budget, as appropriate, for these tasks.

For the Community Gardens project, George organizes meetings around pizza and pot luck dinners. He is a charismatic person and communicates his appreciation of everyone's contributions.

12.3.1 Checkpoint

Working with your team, develop a list of team-building activities. Write these down and add them to the list of tasks developed earlier in your project. Make sure to add time to the project schedule and money to the project budget for these activities.

12.4 Coordination of Work: Sorting Out Team Roles and Responsibilities

Confusion over team roles and responsibilities causes significant problems on projects. There are strategies for allocating work to team members and for sorting out the relationships among team members. A responsibility assignment matrix (RAM) documents the relationships among team members and project deliverables. Organization charts document the reporting relationships among team members and the larger stakeholder community.

12.4.1 Responsibility Assignment Matrices

There are several flavors of RAM. The project manager creates a RAM to document the relationship between team members and project documents and a separate RAM to document the relationship between team members and project deliverables.

Table 12.1 illustrates the RAM for project documents.

When the project manager formats the information in a matrix, he easily sees that there is a relationship between person 1 and the scope statement, person 2 and the budget, and person 3 and the charter.

Table 12.1 RAM for Team Members and Project Documents

	PROJECT CHARTER	SCOPE STATEMENT	SCHEDULE	BUDGET
Person 1		X		
Person 2				X
Person 3	X			
Person 4				

He can also see that no one has responsibility for the schedule, and person 4 is not assigned to any document. Team members easily determine their responsibilities, the project manager sees what work is not likely to be completed because no one is assigned to do the work. He also sees if there are people on the project that have no work assignments.

While this matrix helps determine any oversights in work assignments, it doesn't specify the nature of the relationship between the person and the document. Instead of using X in the cells, the project manager creates a legend that better describes the relationship. For example, he might use the following:

C—Stakeholder is a subject matter expert and responsible as a *consultant* or advisor.
I—Stakeholder is *informed* of the document or activity but does not provide anything to the effort.
R—Stakeholder is *responsible* for doing the activity or writing the document.
S—Stakeholder is responsible for *sign off* on the document or activity.

Table 12.2 illustrates the relationship between the team members and project documents.

Looking over the matrix in Table 12.2, the project manager realizes that no one has "R" for the charter and the schedule. Since "R" denotes person doing the work, the absence of R for any document means no one is doing the work. So, the project manager easily determines if there are any omissions in assignments. The stakeholders see the work for which they are responsible. In addition to a clear definition of responsibilities, it helps the stakeholders plan and manage their time. A stakeholder will pay more attention to an activity on which he must sign off, than an activity where his responsibility is

Table 12.2 RAM for Relationship of Team Members and Project Documents

	PROJECT CHARTER	SCOPE STATEMENT	SCHEDULE	BUDGET
Person 1		R		
Person 2				S
Person 3	C			
Person 4			I	R

Notes: C—Stakeholder is a subject matter expert and responsible as a *consultant* or advisor; I—Stakeholder is *informed* of the document or activity but does not have to provide anything to the effort; R—Stakeholder is *responsible* for doing the activity or writing the document; S—Stakeholder is responsible for *sign off* on the document or activity.

"I" or to receive information. Every document and deliverable should have a person responsible for the work and a different person to sign off on the work. The project manager modifies the RAM on Table 12.2 to include an "R" and "S" for every document.

Table 12.3 illustrates the RAM for the deck project documents. Every document must have a person responsible (R) for doing the work and a person signing off (S) and approving the work. Project stakeholders may or may not have responsibilities for the project documents. The town permitting organization, focused on the deliverables of permits and inspections, has no responsibilities for project documents; it does have responsibilities for deliverables as noted on the RAM for deliverables (Table 12.4).

The project manager also documents the responsibility for work packages and project deliverables and the nature of the responsibility as illustrated in Table 12.4. For example, one team member may be responsible for doing the work, another team member may be a subject matter expert and responsible for providing consulting assistance or reviewing the work. Still another team member may be responsible for signing off on the final deliverable. Much as the organization chart illustrates relationships among team members, the matrix illustrates the relationships between the stakeholders and the work packages.

Appendix P contains the RAMs for the Community Gardens project.

12.4.1.1 Checkpoint Create a legend for your project and build two project RAMs. Develop a matrix documenting the relationship between stakeholders and project documents. Identify the

Table 12.3 RAM for Deck Project Documents

	PLANNING PHASE															EXECUTION		
	CHARTER	SCOPE STMT	WBS	NETWORK DIAGRAM	SCHEDULE	BUDGET	LIST RISKS	RANKED RISKS	RISK RESP PLANS	STKHLDR MATRIX	ENGAGE MATRIX	COMM PLAN	RAM DOCS	RAM WBS	ORG CHART	ACTION ITEM LIST	STATUS REPORTS	CHANGE REQUESTS
Project manager	R	R	RS	RS	RS	R	RS	RS	RS	RS	RS	RS	RS	RS	RS	RS	RS	R
Property owner	S	S				S				C							I	S
Renter		I														I		
Abutter																	I	
Town permitting organization																		
Vendors						I												

Notes: C—Stakeholder is a subject matter expert and responsible as a *consultant* or advisor; I—Stakeholder is *informed* of the document or activity but does not have to provide anything to the effort; R—Stakeholder is *responsible* for doing the activity or writing the document; S—Stakeholder is responsible for *sign off* on the document or activity.

Table 12.4 RAM for Deck Project Deliverables

	1.0 MATERIAL & SUPPLIES PURCHASED	2.0 DECK CONSTRUCTED	3.0 SAFETY PERMITS & INSPECTIONS COMPLETE	4.0 PM DOCS & SERVICES COMPLETE
Project manager	R	R	I	R S
Property owner	S	S	R	I
Renters		I		
Abutters		I		
Town permitting organization			C S	
Vendors of building suppliers	C			

Notes: C—Stakeholder is a subject matter expert and responsible as a *consultant* or advisor; I—Stakeholder is *informed* of the document or activity but does not have to provide anything to the effort; R—Stakeholder is *responsible* for doing the activity or writing the document; S—Stakeholder is responsible for *sign off* on the document or activity.

stakeholders responsible for the documents and list the stakeholder role(s) (C, I, R, S) for the document. Similarly, take the work packages from the work breakdown structure developed earlier and create a second matrix. Identify the stakeholders responsible for these work packages and document the stakeholder role(s) (C, I, R, S) for the work package. Share both RAMs with the affected stakeholders to make sure they are in agreement and have the time to contribute to the project at the time specified in the project schedule. These matrices are also excellent tools to communicate project relationships among all stakeholders.

12.4.2 Organization Charts

In addition to documenting and communicating responsibility for project work, the project manager clarifies and documents the relationships among team members and stakeholders. Some team members report to other team members, some are peers and some manage others on the team. An organizational chart is a good technique for illustrating these relationships. Initially, the project manager creates an organization chart illustrating the relationship among different team roles. Later, as named individuals are assigned to these roles, the project manager adds individual names to the job roles listed on the organization chart (Figure 12.1).

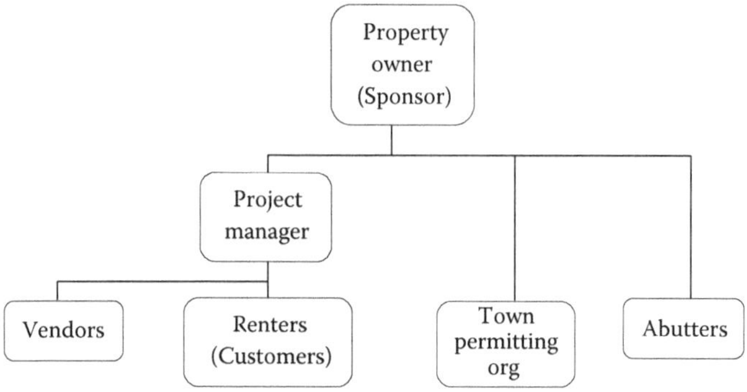

Figure 12.1 Organization chart for deck project.

Since the deck project is small with the project manager also doing most of the work, there is no project team listed on the organization chart. If the project manager enlists the help of other renters in the construction or other work packages, he might list these team members on the organization chart. The renters also have a relationship with the property owner, in that they pay monthly rent, but this is outside the context of the deck project. From the perspective of the deck project, the renters are customers of the completed deck and may be part of the project team that constructs the deck. Since the organization chart is specific to the deck project, it doesn't list a direct relationship between the property owner and the renters.

Appendix Q contains an organizational chart for the Community Gardens case study.

12.4.2.1 Checkpoint Create an organization chart for your project. Discuss reporting relationships with affected team members and obtain agreement. Communicate these organization charts so that affected stakeholders are aware of team structure and roles.

12.5 Next Steps

In the next chapter, we put everything together and complete the project plan.

13

Connect the Dots

Integrating the Project Components

At this point, the project manager has a project plan with sections for charter, scope, schedule, budget, quality, risk, stakeholders, communication, and teaming.

These are all interdependent, and consistency across all this work is important. Since each piece of the plan builds on prior sections, the project plan is likely consistent if the project manager looks backwards to prior pieces of the project plan as he continues development. For example, the tasks of the schedule section build on the work packages of the WBS in the scope section. As the project manager completes the schedule, he looks back to the WBS to see if changes might be necessary. Budget information builds on the tasks; once the budget is complete, he looks back to see if any additions or deletions to tasks might be necessary. Quality builds on the top level deliverables of the WBS, and risk builds on all of these components. As he builds on the information of prior sections, the project manager determines if changes are necessary and ensures he incorporates the impact of these changes as appropriate. For example, if he added a risk response task to the WBS, he verifies that he's added the necessary schedule and budget to the project plan.

Since the risk of not properly connecting all the pieces of the project may be significant, the project manager reviews and validates that all dots are connected. The dots in Figure 13.1 show a picture of a small boat when properly connected. Unfortunately, connecting the dots differently turns the picture from a boat to a boat anchor.

Once the project manager verifies that all the dots are appropriately connected, the project plan is complete and stakeholders sign off, approving the plan. The project manager shares the entire project plan with key stakeholders and explains the significance of the project

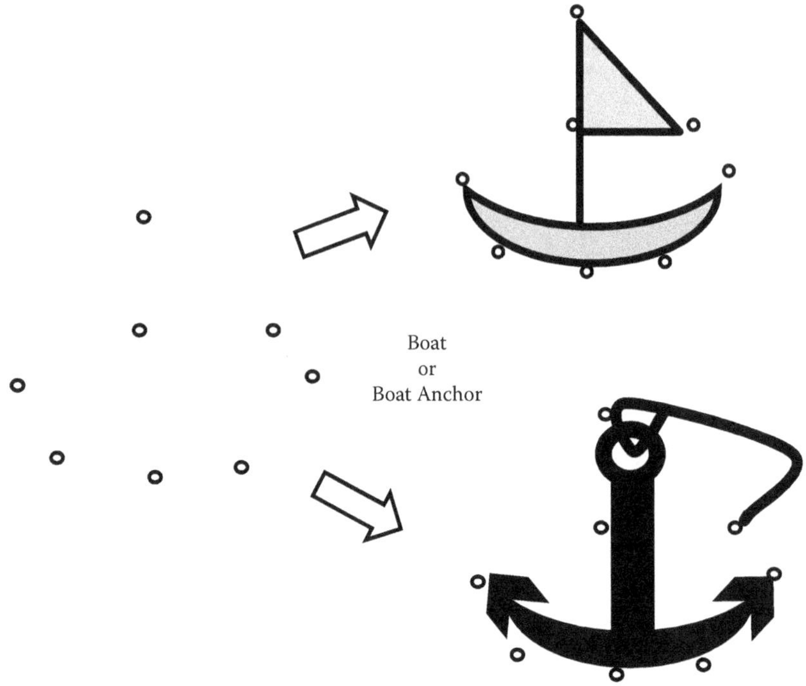

Figure 13.1 Connect the dots.

work and the project deliverables to the stakeholders. Different stakeholders may be more or less interested in different sections of the project plan. The project manager refers to the responsibility assignment matrices (RAMs) and ensures that stakeholders that are responsible (R) or must sign off (S) on sections of the project plan completely understand these sections. He also reiterates the WIIFM (what's in it for me) for the stakeholders.

Once the project manager has an approved plan, he, together with the project team, executes the work of the project—the work defined in the plan.

13.1 How Much Planning Does a Project Need?

It may seem that the project manager and project stakeholders spend a lot of time planning and still have not started working on the project. Have they wasted a lot of time talking and planning? How much time should be spent on project planning?

The time spent planning should be appropriate given the significance of the project; how much will it hurt if the project fails or is incomplete? For a low risk project with limited schedule and budget, the project manager might complete the planning in a matter of hours. For more complex, longer, and more costly projects, planning may take a couple of days. The project manager may have a guideline that he may spend up to 10% of the overall prepared schedule in planning with the remainder of the project schedule for execution.

If the project manager is new to the project process, the planning and the execution will likely take longer. Once the project manager succeeds on a couple of projects, he is familiar with the process and he has templates from earlier projects to use. Both of these shorten the time required for planning.

If the project manager doesn't spend time planning early in the project, it is likely that much of the work executing the project will be wrong and need to be redone. In this situation, the project takes longer, costs more, and likely impacts the relationships among stakeholders.

When the project manager doesn't take the time to plan, he is effectively saying, "I don't have time to plan, but I do have time to do much of the work a second time to get it right." Experience has shown that time spent planning is recovered by accurate, timely, and cost effective execution based on defined and agreed upon expectations and plans.

13.2 Project Plan Document Approvals

An approach to creating awareness of the project pieces as an integrated whole is to organize stakeholder approvals of project plan pieces in a single document. Some of the templates, such as project charter and project scope, have a place for the appropriate approvals in the document itself. Many of the sample tables and matrices do not include a place for approval signatures; the project manager may customize these templates to include stakeholder sign off. In addition to, or in lieu of, approvals on each project document, the project manager captures written approvals on a project plan approval table, which lists all the project documents, the appropriate approver's name and title, approver's signature and date approved. The approver is generally the person with sign off (S) responsibility. The project manager and sponsor complete the first two columns from the responsibility assignment matrix for project documents (Table 12.3 for the Deck project and Table P.1 for the Community Gardens project). The project manager obtains the necessary approvals as the project progresses. Table 13.1 contains a blank project plan approval matrix with suggestions as to who might be appropriate approvers for the documents. The

Table 13.1 Project Plan Approvals

PROJECT NAME			
DOCUMENT	APPROVER NAME & TITLE	APPROVER SIGNATURE	DATE APPROVED
Project charter	Sponsor		
Project scope statement	Project manager Sponsor		
Work breakdown structure	Project manager Owners of top WBS levels		
Extended WBS with tasks	Project manager Owners of the work packages		
Schedule network diagram	Project manager Owners of work packages Task owners		
Project schedule	Project manager Sponsor		
Project budget	Project manager Sponsor		
Project quality plan	Project manager Sponsor Customers		
List of prioritized risks	Project manager Sponsor Key stakeholders		
Risk response plans	Project manager Risk response plan owners		
Stakeholder matrix	Project manager Sponsor		
Stakeholder engagement matrix	Project manager Sponsor Key stakeholders		
Project communication plan	Project manager Key stakeholders		
RAM for documents	Project manager Key stakeholders		
RAM for deliverables	Project manager Key stakeholders		
Organization chart	Project manager Key stakeholders		

project manager approves all documents except the project charter. Theoretically, the project sponsor creates the project charter; in reality, the project manager writes it on behalf of the sponsor. As the person funding the project, the sponsor must approve the rationale for the project as documented in the charter.

As a best practice, the project manager obtains approval from the stakeholders affected by the document. For example, the owners of top WBS levels and work packages may be affected by the organization of the WBS charts; the project manager requests their approval of this work even though they do not have primary responsibility or sign off on these documents. In the case of risk response plans, the owner of the plan varies based on the focus of the plan. The project manager relies on the responsibility assignment matrix for owners and approvers of these plans. The project plan approval table does not include risk categories, probability scales, and impact scales. These are cross-project standards; while used in the current project, they are not necessarily specific to the current project. The project manager may choose to include these in the approval table, but it is not necessary, as they have been approved outside the context of a specific project.

13.3 Project Plan

The final project plan is the compilation of all the documents created by the project manager and project team and approved by appropriate stakeholders. Once this is completed and approved in its entirety, any changes are subject to the change control process.

Table 13.2 shows the sections of the project management plan for projects.

13.3.1 Checkpoint

Review the project plan for your project. Using the RAM or the project plan approval table, schedule time with the appropriate stakeholders to reiterate the relevant information, and obtain their sign off. This may be a series of individual meetings, or meetings with many of the stakeholders at the same time. It may be effective to provide copies of the plan prior to the meeting so the stakeholders have time to read and review the relevant information. If you have involved the

Table 13.2 Project Plan Contents

Project charter
Project scope statement
Work breakdown structure
Extended work breakdown structure with tasks
Schedule network diagram
Project schedule
Project budget
Quality plan
List of risks (both threats and opportunities)
Risk categories
List of top risks prioritized (threats and opportunities) including probability, impact, risk score and ranking
Risk response plans for top priority risks (both threats and opportunities)
Stakeholder matrix
Stakeholder engagement matrix
Project communication plan
Responsibility assignment matrices for project documentation and project deliverables
Project organization chart
Project plan approvals

stakeholders in the development of the project plan, as suggested earlier, the stakeholders are already familiar with the project plan information, and this is a final review prior to the sign off. If there are additional documents for your project, add them to the project plan approval table and obtain the appropriate sign offs.

13.4 Next Steps

The next chapter contains a high level approach to project execution, using the deck project as an example. The final chapter contains a step-by-step process for project execution.

14
After the Plan, We Execute

When the project manager commits to deliver a project, he is really committing to deliver a product and a *process* to ensure the accurate and cost-effective delivery of that product.

The project documents (charter, scope, WBS, schedule network diagram, budget, quality plan, risk plans) tell the project team and stakeholders what should be done; the team now executes these plans. The stakeholder information, communication plans, resources, and teaming information guide the implementation of the work to be done as specified in the scope statement, quality plan, schedule, and budget. The toolkit lists the documents, standards, tactics, and processes to structure the work of the project.

Once the project plan is in place, the project team and stakeholders implement the plan to accomplish the work of the project. The project manager holds meetings, engages resources, and monitors progress against the project schedule and budget. If there are variances of actual performance relative to the plan, the project manager influences the team to accomplish more so that the actual results better conform to the plan. Alternatively, the project manager looks to adjust scope, schedule, budget, or quality depending on the relative importance of each of these to the project. If scope and quality are most important, the project manager adds more resources which increase the budget. If budget is the top priority, the project manager reduces scope. This would require a change to the project plan, following the documented change control procedures and approvals. The goal is to discover variances soon enough to proactively take action to bring the project back on plan.

14.1 The Product of the Project

In addition to the project plan documents organized into the project plan, there is the product of the project. The deck is the product of the deck project.

The process for managing the product of the project and the project itself is similar; the project manager and team apply the process to different deliverables. One deliverable is a successful project plan; the other deliverable is a quality product.

14.2 Execution of the Deck Project

This is a relatively small project from the perspective of budget and number of resources involved. It is somewhat unique, in that the project manager is also the main resource working on the execution of the plan and the creation of the product of the project. The project manager may enlist the assistance of his family or friends who rent the house with him and are beneficiaries of the deck. If he creates a project team, he reassigns tasks to his team and changes the project schedule, RAMs, risk plans, communication plan, and teaming approaches as appropriate. He obtains the consensus of these team members and approval of the property owner as the project sponsor.

A key milestone is securing the building permit. Since the town requires that the property owner complete and submit this application, the project manager assigned this task to the property owner. Since this approval is necessary for work to begin, the project manager closely monitors the progress of this task. Once the project has this approval, the project manager begins to acquire supplies and implements the risk response plan for the threat of building inexperience.

While sourcing supplies for the deck, the project manager discovers that prefinished, weather resistant wood planks are available for a short time at the local hardware store at a significantly reduced price. Analysis suggests that this potential change is beneficial to the project; the project manager documents his analysis of this potential change in the change request and analysis form (Figure 14.1) and obtains the project sponsor approval. He then updates the appropriate project documents, including scope statement, schedule, schedule network diagram, budget, and quality documents.

Change Request

Change Name: Use Pretreated Planks **Date Required:** ASAP
Date Submitted: **Category:** Scope and Budget **Requested by:**

Description of Change: The local hardware store is selling pretreated wood at a greatly reduced price. When the team considered this as a positive risk, the risk score did not support inclusion of these materials in the project scope. The reduced cost now makes this an attractive opportunity. This wood eliminates the need for weatherproofing and finishing the wood. The sale on the wood ends Friday, so the team needs approval to proceed ASAP.

Benefits

1. Complete project 2 days early
2. No need for periodic painting and maintenance of the deck
3. Environmentally friendly

Limitations

Net additional cost of $200. Save 20% of $500 cost of paint (save $100); increase cost of wood by 20% or $300 (from $1,500 to $1,800). Reduced maintenance costs are another cost benefit. (The budget listed $2,500 for deck materials; this included $1,500 for wood, $500 for paint, and $500 for other supplies.)

Impacts

Scope: use pretreated wood with positive impact on long-term maintenance

Schedule: save 2 days

Budget: net additional cost of $200

Quality: positive impact on environment and aesthetics

Risks: no impact

Recommendation

Recommend that the team implement this change and purchase the pretreated wood at the sale price. Change approved.

Reviewers and Date: ____________________
Approvers and Date: ____________________

Figure 14.1 Change request and analysis for deck project.

To implement the risk response plan (Figure 7.1), he schedules the class on building a deck offered by the local hardware store for the resource assigned to this task. The project manager might implement these tasks while waiting for the permit approval; the risk is that this effort is wasted if the town denies the building permit.

The property owner is responsible for managing the concerns of the neighbors whose property abuts the property for the deck. Since this has the potential for long-term negative consequences if not addressed positively and promptly, the project manager monitors the efforts of the property owner in this area. The project manager implements the communication plan, including weekly status reports, status of permits and inspections, orders for materials, and vendor deliveries. He, or his designee, orders the building materials and constructs the deck. He compares budget and spending and promptly reports significant variances to the property owner; he obtains approvals for additional spending, if necessary. He also monitors the schedule and reports significant variances.

Although the approved budget of 92 labor hours and $2925 compared favorably with the estimates in the project charter ($3000 for supplies; no estimate for labor), the approved change request introduced an additional cost of $200. This 7% increase over the approved $2925 cost for materials is well within the allowable variance of plus or minus 10%.

Once the project manager completes the deck, he meets with key stakeholders. This includes the property owner and the team members that helped with construction, supplies acquisition, and management of stakeholders such as the abutters, vendors, and town building departments. They review the project overall, identify what worked well and what didn't work well, and create lists of things to do, and not do, on future projects. This is often called a "lessons-learned" meeting. Figure 14.2 illustrates the outcome of the lessons-learned meeting for the deck project.

Lessons Learned

Summary of Deck Project

1. Project within allowable budget variance—spent $3125 against a plan of $2925. This is within the allowable variance and the sponsor supports the change request that caused the additional spending.
2. Project delivered late by 1 week—delay due to more bad weather than anticipated and late vendor delivery of supplies.
3. Abutters experienced little disruption during deck construction (as reported by abutters).
4. Property owner pleased with aesthetics of deck (as reported by property owner).

What Worked Well

1. Communication plan ensured that stakeholders were aware of project status and issues.
2. Property owner, responsible for permitting and managing expectations of abutters, positively reinforced these important relationships.

What Did Not Work Well

1. Vendors did not prioritize delivery of supplies—often arrived late negatively impacting the schedule.
2. Difficult to motivate team; the WIIFM of having a deck for outdoor activities did not incent on time completion of tasks. Need a WIIFM that better motivates the team members.

Suggestions

Consider paying construction workers to obtain expertise and on-time completion.

Figure 14.2 Lessons learned for deck project.

Table 14.1 Quality Plan for Deck Project—Post-Project Completion

Priority [1 = high, 2 = medium, 3 = low]: 1-Scope; 3-Schedule; 2-Budget				
PROJECT METRICS				
CATEGORY	DESCRIPTION—PROJECT CAPABILITIES	TARGET METRIC	ALLOWABLE VARIANCE	ACTUAL PERFORMANCE
Scope	Project meets purpose as defined in charter	Sign off and approval by sponsor and customer (property owner and renters)	None	Met target metric
Schedule	Project meets schedule constraints	Schedule milestones as defined in project schedule	±10% or as defined in schedule	City workers staged work slowdown during contract dispute; permit delay of 1 week
Budget	Project meets budget constraints	Costs incurred as defined in project budget	±10% or as defined in budget	Costs in line with allowable variance. Sponsor supports additional spending
PRODUCT METRICS				
CATEGORY	DESCRIPTION—PRODUCT FEATURES	TARGET METRIC	ALLOWABLE VARIANCE	ACTUAL PERFORMANCE
Regulatory	Deck meets all building codes	Obtain building permit	No acceptable variance	Met target metric
Safety	Deck supports expected load of people, equipment and furniture	Supports combined weight of 2000 pounds	±10% metric acceptable	Exceeds target metric; city approves for 2500 pounds
Aesthetics	Deck complements look and feel of house	Owner satisfied Renters satisfied	Not applicable	Both owners and renters sign off on look and feel
Usage	Deck in use for at least 3 seasons	Renters report using deck April through October	Renters report actual use limited to May through Sept.	Unable to use deck regularly due to sun exposure, bugs, and temperature variations

The project manager returns to the quality metrics table for the deck project and completes the rightmost column to capture quality information only available once the project finishes. Overall, the project was successful meeting the product metrics, within acceptable variances for schedule delay and securing sponsor approval for overspending (Table 14.1).

Appendix R contains Project Execution for the Community Gardens project.

14.3 Next Steps

The final chapter lists the steps for project execution.

15 Steps to Execute Your Project

If you, as the project manager, have been working on your project while reading this book, you have a completed project plan, reviewed and approved by stakeholders. The project manager now begins to execute this plan. If there is no project plan, now is the time to create one with your stakeholders *before* beginning project execution.

Below are the steps for project execution along with some problems that may occur if there is no project plan. Recall that project execution includes both the steps to create the project plan and the steps to execute this plan. Each step is a predecessor for the steps that follow, so it should be complete before proceeding. Sometimes, the project manager decides to continue to the next step, even though the current step is not yet complete, with the understanding that the current step will be completed by a specified future date.

The steps below do not strictly follow the sequence of project core, toolkit, and project context. As the project manager goes through the project planning and execution, the actual sequence of steps may be different. The project manager customizes the sequence based on the unique characteristics of the project, the project stakeholders, and the project context.

15.1 Step 1: Create a Project Charter

Before the project manager completes the project charter, he talks with the project sponsor and they develop a mutual understanding of the project processes and the project deliverables. It is important that he develops a complete big picture of the product of the project and the high level project processes. From this information, the project manager and the sponsor determine project parameters such as schedule, budget, quality, and risks. The big picture includes the project objectives and success criteria, along with a high level project description. The focus at this point is the "what" of the project which the project manager later expands into the project scope. The project parameters will likely change as the project manager clarifies and elaborates the project scope. The charter is the foundation and the basis for all other project decisions and documentation. If the project basis changes significantly, it may be better to close out the existing project and start over with another project and project charter. The project manager continues with the existing charter if the scope, schedule, budget, or quality change, but the project objectives and success criteria, along with the high level project description and boundaries remain the same. As long as the project "what" does not change, the project manager continues building on the project charter, making changes to the project parameters as necessary.

The project charter has a section entitled "Description and Boundaries." This section of the charter addresses both the project processes and a description of the product of the project. The boundary between project and product often blurs, and it is fine to include product information. However, it is easy to focus on the product to the exclusion of the project processes; it is important not to focus on the product information to the exclusion of the project processes.

If the same person is both the sponsor and the project manager, it is still necessary to document the project charter. The charter is the basis for all other project documentation and a key vehicle for communicating the project rationale to other stakeholders. Verbal information is often forgotten and subject to misunderstanding and variation over time. Writing this information forces the project manager to think through the project; the written word often looks very different than unspoken thoughts about the project.

The project manager stores the project charter so that it is available to other project stakeholders. If the project manager is working in a nonelectronic environment, he may keep the project documents in a room or file cabinet that serves as project headquarters. In an electronic environment, this shared space may be a dropbox, shared folder, or Facebook page, or any other easy to access location. The project manager calls this the project document repository.

15.1.1 Checkpoint for Project Status

This step is complete when there is a written and approved project charter.

15.1.2 Risks of No Written Project Charter

If there is no written project charter, the project manager may plan the wrong project or product. The deck project charter refers to the project as "Deck for Rental Property." However, the charter notes that railings are out-of-scope for the deck, suggesting that perhaps the real intent of the project is a patio that is level with the ground, rather than a raised deck. Absent the documentation and specification of out-of-scope information, the project manager might plan and build a raised structure rather than a structure level with the ground.

15.2 Step 2: Create the Project Scope Statement

If substantial time has elapsed since the project manager completed the charter, he reviews the charter information with the sponsor and notes any information that has changed. The project manager and the sponsor identify the information necessary for the project scope statement. The project manager then schedules meetings with the stakeholders listed in the project charter. These meetings may be with a single stakeholder or may include several stakeholders; the meetings may be face-to-face or via telephone, Skype, or any other tool that allows for synchronous communication.

Synchronous communication requires that everyone participate in the conversation at the same time. Asynchronous communication includes conversations where the participants are not in the conversation at the same time. For example, e-mail communication and text messaging are asynchronous; one party sends a message when it is convenient, and the other party reads and responds to the message when convenient, which may be several hours or days later.

The purpose of these conversations is twofold. The project manager shares information and also builds a relationship with the stakeholders. Face-to-face and synchronous communication are most effective at both of these tasks.

The conversations address several topics. The project manager reviews the project charter and ensures that the stakeholders support the project purpose, objectives, success criteria, and requirements, and they understand the top level information on risks, schedule, budget, assumptions, and constraints. The project manager solicits the stakeholders' perspective on the contents of the project scope statement. If key stakeholders do not support the project, the project manager and sponsor decide if project success is possible without this support; if not, they must either positively engage the stakeholders or cancel the project.

The final scope statement includes information from the project manager, project sponsor, and project stakeholders. The project manager asks each stakeholder if there are other stakeholders, not yet listed, that the project manager should include on the list of key stakeholders. In this way, the project manager discovers other people who might be affected by the project early in the process and lists the stakeholders on the stakeholder matrix.

Once the project manager has consulted all the stakeholders, he completes the project scope statement and reviews it with the project sponsor. He then shares it with the stakeholders that have contributed to the scope content and obtains the necessary approvals and sign offs. He stores the approved document in the project document repository.

15.2.1 Checkpoint for Project Status

This step is complete when there is a written and approved project scope statement.

15.2.2 Risks of No Written Scope Statement

As noted above in project charter, writing the scope information forces the project manager to think through and clarify information. Stakeholders have access to the same information, and the information is not subject to the risks of forgetting what was agreed to after some time has passed.

If the project manager fails to document the project scope, the WBS and other project information based on the project scope may be incomplete, partially incorrect or totally wrong. Errors introduced early in the project life are very costly; they have a cascading effect on all the project information and decisions that follow. For this reason, it is especially important that the information developed early in the project is accurate, documented, and agreed to by key stakeholders.

15.3 Step 3: Create Stakeholder Matrix

The project manager starts the creation of the stakeholder matrix with a list of project stakeholders from the project charter and from his work developing the scope statement. If the list of stakeholders is long, the project manager may complete the necessary information only for the top-ranked stakeholders. However, it is best practice to list the less influential stakeholders at the end of the stakeholder matrix or as a separate watch list. If these less influential stakeholders later become more significant, the project manager then completes the information in the stakeholder matrix, and updates the stakeholder engagement matrix and the communication plan as appropriate. The project manager periodically reviews this watch list with key stakeholders.

As the project manager develops the stakeholder matrix, he shares information about what the project needs from the stakeholders and asks the stakeholder what he needs from the project. The project manager records this information in the stakeholder matrix. Working with key stakeholders, the project manager groups the stakeholders into the appropriate power (high or low) and interest (high or low) categories and assigns priority rank based on the power and interest categorizations. These ratings may be politically sensitive, with a stakeholder claiming high power, when the project manager and sponsor perceive the stakeholder to have low power over the project. The project manager demonstrates discretion and care in assigning and communicating these rankings. In some cases, it may be best to modify the documented ranking to keep the stakeholder satisfied.

As the project manager continues project planning and execution, he may discover new stakeholders, and rankings may need to be adjusted. The stakeholder matrix is a working document, subject to change until signed and approved just prior to project execution. Changes after approval go through the change control process.

15.3.1 Checkpoint for Project Status

This step is complete when there is a written and approved stakeholder matrix.

15.3.2 Risks of No Written Stakeholder Matrix

There are only so many hours in a day, and a project manager must prioritize his many responsibilities and tasks to ensure that the most important get the necessary attention. Similarly, on many projects, there are more stakeholders than the project manager has the time to manage. If there is no stakeholder matrix, the project manager may not realize the stakeholders' expectations of the project, and the stakeholders may not realize what the project needs from them. Absent this information, task assignments and responsibilities for the project schedule, communication plan, and teaming efforts may be incomplete or inaccurate. Further, the project manager may spend time with stakeholders that have little influence over the project and ignore stakeholders that make the project successful.

15.4 Step 4: Create Stakeholder Engagement Matrix

Building on the stakeholder matrix, the project manager, working with key stakeholders, identifies the current and preferred level of engagement for the top-ranked stakeholders. For stakeholders with a rank of 1 or 2, the project manager documents his plan to move the stakeholders from today's level of engagement to the preferred level of engagement and implements this strategy as the project progresses.

15.4.1 Checkpoint for Project Status

This step is complete when there is a written and approved stakeholder engagement matrix.

15.4.2 Risks of No Written Stakeholder Engagement Matrix

Project success is challenging when key stakeholders are not appropriately supporting the project. Aligning the current and preferred levels of stakeholder engagement provides an opportunity to reposition negative and neutral stakeholders to a more positive project bias. Without a stakeholder engagement matrix and plan to reposition stakeholder bias, the project manager may miss opportunities to create a more supportive project environment.

15.5 Step 5: Create the Communication Plan

Building on the list of stakeholders from the stakeholder matrix and the stakeholder engagement matrix, the project manager completes the communication information required for each stakeholder. The project manager has each stakeholder validate the information and agree to the role as approver, reviewer, or executor for the information. As the project acquires new stakeholders, the project manager updates the communication plan.

15.5.1 Checkpoint for Project Status

This step is complete when there is a written and approved communication plan.

15.5.2 Risks of No Written Communication Plan

If there is no communication plan, information may not be shared among stakeholders in a timely manner. Stakeholders may not set aside the time necessary to approve, review, or execute documents or tasks, and may not understand what information the project needs from them. Communication is a critical success factor for projects and project managers.

15.6 Step 6: Teaming, Create RAM for Documents, Create Organization Chart

At this point, the project manager has a preliminary list of project stakeholders and begins team building and the identification of team relationships. The project manager lists the team building activities for the project team, and if they are significant in terms of time or cost, he includes them in the project schedule and budget. Even if they are insignificant in terms of time, the project manager lists them on the project schedule to insure that they are not forgotten. He creates the

RAM for project documents, assigning team member responsibility for project documents. Since he has not yet developed the WBS and the WBS work packages, he cannot create the RAM for work packages; this happens as part of the decomposition of the top level WBS into the work packages. Once he completes the RAM for project documents, he can also create the project plan approvals table.

The project manager creates an organization chart for team roles; if the team members are assigned to these roles, he can also complete the organization chart for team members at this time. Otherwise, he completes the organization chart for team members over time as team members are assigned to the various roles.

The project manager may develop these project documents alone or may include the project sponsor and other key stakeholders. In either case, he reviews these documents with key project stakeholders to ensure that they support these efforts.

15.6.1 Checkpoint for Project Status

This step is complete when there is a list of teaming activities, a RAM for project documents, and a project organization chart finalized and shared with key stakeholders.

15.6.2 Risks of No Written List of Teaming Activities, RAM for Project Documents and Project Organization Chart

Failure to coalesce the core stakeholders into a functioning team impedes timely and cost-effective project completion. The organization chart ensures that team members understand their relationships to each other and among the team members, so there is no confusion as to team roles. The RAM for project documents ensures that all work is properly assigned, and everyone knows their responsibilities for project documents.

15.7 Step 7: Create the Work Breakdown Structure (WBS), RAM for WBS Work Packages

The project manager must determine how to structure the WBS. The WBS may be organized by deliverables or by the chronology of project phases. The project manager then determines which stakeholders should be involved in the development of the WBS. The purpose of the WBS creation is twofold. It defines the "what" of the project in the form of deliverables, and the project manager uses the process of developing the WBS to create shared ownership of the project deliverables. The project manager encourages key stakeholders support and ownership of the project.

The project manager schedules a meeting for WBS development and invites key stakeholders. He reviews the project charter and scope statement at the meeting. Using sticky notes and a flip chart (or white board), the project manager guides the stakeholders in the creation of the WBS. Once the stakeholders understand the use of sticky notes and the top level WBS deliverables, the project manager steps back and allows the stakeholders to collaboratively build the WBS. The project manager monitors the stakeholder progress and guides the effort if the stakeholders stray from the project "what" and begin to identify the project "how" (or tasks). If stakeholders get too detailed, the project manager works with them to move the sticky notes around so that there is reasonable organization of top level deliverables. Once the stakeholders agree that the WBS is complete, the project manager numbers the deliverables and work packages and takes a photo of the WBS. He keeps the flip chart paper with the sticky notes and may e-mail the photo to the key stakeholders as documentation of the group effort. If the project manager uses a word processing or other computer application, he may type the WBS information into this program, creating a WBS in an indented hierarchy format. He may also create a typed graphical version of the sticky note WBS.

The project manager assigns responsibilities for the project deliverables and completes the RAM for WBS work packages.

15.7.1 Checkpoint for Project Status

This step is complete when there is a completed and approved WBS and RAM for WBS work packages.

15.7.2 Risks of Not Creating a WBS

The WBS is the single most important document of the project plan. It is the basis for the project schedule and budget; it identifies the work to be done to create the project deliverables and the product of the project. If the project manager does not create a WBS, there is a risk that the project deliverables and the product of the project are not clearly defined and agreed to by the key stakeholders. There is also the risk that the stakeholders view the project as the project manager's project, rather than a project owned by all key stakeholders. In this situation, the stakeholders may not participate positively in the project work, and they may fail to encourage their colleagues to support the project work.

15.8 Step 8: Create the Project Schedule

The project schedule consists of the WBS work packages extended to include the tasks necessary for each work package, the schedule network diagram, and the project schedule.

15.8.1 WBS Extended to Include Tasks

If some time has passed since the project manager last reviewed the WBS, it is good practice to look over the scope statement and WBS and address any questions that surface. The next step is to extend the WBS with the tasks necessary to create the project deliverables, continuing the use of sticky notes but using a different color or shape to differentiate the work packages and the tasks. The project manager augments the project scope, or "what," with the "how" in the form of tasks to complete the work of the project. It is critically important to involve the people who will actually do the work of the project. If one person identifies the tasks and the time necessary to complete the tasks, and the project manager then assigns the tasks to another person, there is the risk that the person now responsible for the tasks may say that the tasks are not appropriate or that the times allocated to these tasks are not reasonable.

Ideally, the project manager identifies the people responsible for the deliverables and instructs them to list the tasks necessary to develop the deliverables. If the team members have not previously worked with WBS and WBS extended with tasks, the project manager may work with one or several team members on this. Once team members have worked on the decomposition of deliverables into tasks, they may mentor other team members who have not done this type of work in the past.

Sometimes, the project manager may not know who will be doing the project work. In this situation, the project manager assigns the work of decomposing the work packages to a potential team member who has the necessary skills and knowledge to identify the tasks necessary to create the work package.

Once all of the work packages have been decomposed into the appropriate tasks, the project manager, together with the team members who decomposed the work packages, reviews all the tasks across all the work packages to discover overlaps or omissions that may not be apparent when looking at only one of the work packages. Using sticky notes makes it easy to add, delete, or move tasks from one work package to another. It's not necessary that the team create a "perfect" set of tasks at this point; once the team proceeds to sequence the tasks to create the schedule network diagram, overlaps or omissions generally surface, and corrections are made to the extended WBS at that time. The project manager extends the numbering of the work packages to the tasks and takes a photo of the extended WBS.

If there are many tasks for a work package, or if the people responsible for the work package and the tasks are different, the project manager may develop a RAM that shows the relationship between the stakeholders and the tasks, identifying the tasks by number and name.

15.8.2 Schedule Network Diagram

The project manager brings the team members who developed the tasks for all of the work together to sequence the tasks. The WBS extended with tasks includes the deliverables at the top level, the work packages at the bottom level of the WBS, and the tasks associated with each work package organized beneath each work package.

The sequence of the tasks is independent of the task organization by work packages which are organized by deliverables. The project may begin with a task from deliverable 3, followed by tasks from deliverables 1 and 7. The numbering of the tasks ensures that task association with a particular work package is clear. For example, task 7.1.3 belongs with deliverable 7 and work package 7.1. The team members doing the work make sure that all necessary prerequisites are in place.

Rather than taking the task sticky notes from the extended WBS, the project manager replicates these sticky notes to build the schedule network diagram. The original extended WBS is available for reference and changes at a later time. As the team sequences the tasks, they may find that additional tasks are needed or some tasks are unnecessary. The team revises the extended WBS as appropriate. Once the team agrees that the task sequencing is correct and complete, the project manager takes a photo of the schedule network diagram to record and preserve the work. He shares this with the project sponsor and other key stakeholders. He may also have the team review the schedule network diagram after a week or two has passed, as sometimes the team sees the sequencing differently with the perspective of time.

Once the project team identifies the sequence of project tasks, the individual task owners determine how long it will take to complete the task. As noted earlier, the task owner identifies both effort and any delays or wait time (such as delay due to waiting for supplies to be delivered). The task owner identifies any materials or supplies that he may require to complete the task, including quantities and when required.

15.8.3 Project Schedule

While the task owners identify the time to complete the task(s), the project manager creates the outline for the project schedule. Using a computer program like word or excel, or pencil and paper (lined or graph paper), the project manager lists the tasks along the left side in order based on the sequence in the schedule network diagram, leaving columns for applied time, % time, duration, and number of resources. Then, he creates columns for each day or week (whatever unit of measure used for the project).

For each task, he enters the information for each column and denotes work time with an "x" and delay time with "d" to generate an accurate schedule plan. For each task, the project manager allows a row for people resources and a second row for material or services to be purchased from outside the project. The project manager completes this level of detail later when preparing the budget. The task owner specifies material resources and services. The project manager may attach these specifications at the end of the project schedule or may include them separately in the project plan. In either case, these requirements must contain the number of the task for which they are needed.

The project manager makes sure that the resources are available for the times the tasks are scheduled and may schedule a task for a different week if the resource is not available. The project manager also considers other schedule constraints such as school vacations, holidays, possible nonwork time due to poor weather in the form of storms, excessive heat or cold (if project requires work outdoors), and limited availability of team members due to other job and family commitments.

The project manager compares this projected schedule with the schedule constraints from the project charter and ensures that the project can be completed within the allotted time. If it appears that the project will not be complete by the schedule due date, the project manager and team revisit and modify task sequencing, determine if there are opportunities to complete tasks more quickly by applying more resources or tools and possibly reduce scope to complete the project by the required date. Any changes require that the project manager adjust the appropriate project documents such as WBS, tasks, resources, or time to complete tasks.

At this point, there is a project schedule that forms a plan to complete the project by the date specified in the project charter.

15.8.4 Checkpoint for Project Status

This step is complete when there is a completed and approved WBS extended with project tasks, schedule network diagram, project schedule, and updates to the RAM for WBS work packages, and RAM for project tasks, if necessary.

15.8.5 Risk of Not Creating an Extended WBS, Schedule Network Diagram, and Project Schedule

If the project manager and team fail to identify the tasks necessary to develop each work package, there is the risk that the work to complete the project will be underestimated, with the project likely late and over budget. The lack of a schedule network diagram may result in team members sitting idle waiting for equipment that was not ordered with sufficient lead time. Alternatively, the equipment may be ordered too soon and subject to pilfering while sitting around. Another risk is that team members will work on the tasks that they most enjoy, so some work will be complete early, while other tasks will happen later, possibly delaying the work of other team members. The risk of no schedule includes schedule delays due to failure to consider nonwork time and the inability to assess the impact if a task should start or finish early or late.

15.9 Step 9: Create the Project Budget

Once the project schedule is complete, it is relatively easy to complete the project budget. The project manager identifies all goods and services to be purchased for the project from the row entitled "Material" for each task and enters the cost of the materials. If the materials or services are very technical, it may be necessary for the task owner to obtain the cost estimates. If they are commodity items and the description and quantities are clear, it may be possible for other team members to obtain the costs. Techniques to estimate the cost of these items include searching on web sites such as Amazon, calling vendors, or talking to others who have recently purchased similar items or services.

If the people working on the project are paid, the project manager includes the cost of these labor resources. If the project manager doesn't know the salary for the resource or if the information is confidential, he estimates these costs. There are many web sites with salary ranges for different categories of workers. The project manager may call a local job placement agency or state Department of Employment for salaries for different categories of workers.

Once the project manager completes this information, he sums the costs and has a budget for different time periods (quarterly, yearly) and for the project overall. He compares this proposed budget against the budget constraints in the project charter. If there are significant variations, he consults with the project sponsor to see if additional funds may be available. If not, the project manager considers purchasing lower quality materials or reducing scope. In either case, the project manager updates project documents including scope statement, WBS, project schedule, and quality plan. If the project costs are significantly less than originally forecast in the project charter, this may be a red flag that the team has failed to include all project deliverables in the estimates. The project manager reviews the WBS and WBS extended with tasks to make sure that nothing has been missed in the schedule and budget processes.

15.9.1 Checkpoint for Project Status

This step is complete when there is a completed and approved project budget.

15.9.2 Risk of Not Completing a Project Budget

If the project manager doesn't complete the project budget, he doesn't have a realistic estimate of what the project will cost, and when the costs will be incurred. The risk is that the project will cost more than estimated or that the project will be incomplete due to insufficient funds.

15.10 Step 10: Create the Quality Plan

The project manager, sponsor, and key stakeholders define the quality characteristics of the project processes and the product of the project. The project manager reviews the project scope statement and the WBS to identify relevant quality categories and quality measures for the project and product of the project. He then reviews these with the sponsor and key stakeholders to obtain their agreement that the list of quality goals is accurate and complete, and he works with them to identify the target metrics and the allowable variances. While concrete, objective, and measureable quality measures are best, it is not always possible to quantify quality measures. It is preferable to develop subjective quality measures rather than omit a relevant quality measure.

Once the project manager completes the quality plan matrix, he shares it with all appropriate stakeholders and obtains approvals.

15.10.1 Checkpoint for Project Status

This step is complete when there is a completed and approved quality plan.

15.10.2 Risk of Not Completing a Quality Plan

The quality plan documents the stakeholder consensus as to what is an acceptable product or process. If there is no plan or it is not documented, there may be disagreement as to what quality measures and metrics are appropriate for the project. The team may deliver a product that does not meet the expectations of the sponsor or customer, resulting in nonpayment of expenses, demands that the product be reworked, or outright rejection of the product of the project.

15.11 Step 11: Identify Risks

Risks are best identified by a wide cross section of project stakeholders. The project manager has a preliminary list of stakeholders from the project charter, and likely has identified additional stakeholders while developing the scope statement, schedule, budget, and quality plan. If the project manager has followed the best practice of listing these stakeholders on the stakeholder matrix, he has a list of stakeholders from which to choose.

The project needs a list of risk categories and scales for probability and impact. Absent any other lists, the project manager uses the lists identified for the deck project and the Community Gardens project and adds or deletes from these lists and scales as appropriate. He then selects the key stakeholders and schedules one or more meetings preferably with everyone physically present. If stakeholders are unable to participate in a face-to-face meeting, the project manager may meet with individuals or small groups, sharing and integrating the information as appropriate.

The project manager uses brainstorming or sticky notes for the team to generate a list of positive and negative risks. These same tools work to assign probability and impact scores; the risk score is calculated from the probability and impact values. The project manager may prepare the risk response plans or he may delegate this work to a subject matter expert or other stakeholder knowledgeable about the risk. He may also collect the information from the subject matter expert and then write the risk response plan himself.

The project manager reviews the risk response plans with the project sponsor and a subset of project stakeholders; it is often not necessary to include all stakeholders in the review process. Once the risk response plans are approved, the project manager adds these risk response plans to the WBS, as deliverables or work packages. He assigns this work to the appropriate team member for estimating or validating time and effort and includes them in the project schedule, budget, and quality plan.

15.11.1 Checkpoint for Project Status

This step is complete when there is a completed and approved list of risk categories, scales for probability and impact, list of project risks, risk scores, risk response plans, and updates to the WBS based on risk response plans.

15.11.2 Risk of Not Doing a Risk Plan

It is best practice to validate the project scope, schedule, and budget by asking "what can go wrong or what are the upside opportunities" and "how likely are these problems and opportunities." If the project manager and project team do not assess upside opportunities and downside threats, they may miss these events, to the detriment of the project. Generally, failure to complete a risk assessment and plan results in negative variances for scope, schedule, budget, and quality.

15.12 Step 12: Publish the Project Plan

Throughout the planning process, the project manager communicates the plans and secures the necessary approvals. As a final step, he reviews the plan in its entirety to make sure all the "dots are connected," and the pieces of the plan appropriately integrate with each other. For example, he ensures that any changes have been included in the WBS, and the project schedule and budget reflect these updates. The project manager and project sponsor approve the final plan, and the project manager ensures that the current project plan is available to all project stakeholders. The project manager obtains approvals from other stakeholders, as documented in the project plan approval table.

15.12.1 Checkpoint for Project Status

This step is complete when there is a completed and approved project plan available to project stakeholders.

15.12.2 Risk of Not Integrating and Publishing the Project Plan

The failure to review all pieces of the project plan may result in project disconnects. For example, the project may have tasks without corresponding schedule and budget, risk response plans may lack corresponding time and budget, there may be work with no owner, or there may be significant imbalances in the work load with some team members with too much work, and other team members sitting idle.

15.13 Step 13: Execution

The project manager meets with the project sponsor and begins project execution with a kick off meeting. He follows the project plan, implementing the communication plan, and managing the team effort to execute the project tasks as defined in the project schedule. He stays in touch with the team members and key stakeholders using the communication techniques described including MBWA (Managing by Walking Around). The project manager stays alert to any team or stakeholder dissension and manages conflicts as they occur. Once the team completes the project deliverables and the project sponsor and customer sign off on the deliverables, the project manager holds the lessons learned meeting, and everyone moves on to the next project.

15.13.1 Checkpoint for Project Status

This step is complete when all project and product deliverables are completed and approved as defined in the project plan. The project manager updates the project plan with approved variances.

15.13.2 Risk of Not Executing According to the Project Plan

If the project manager doesn't follow the project plan as approved and doesn't keep the project plan current as changes occur, the project is at risk of incomplete or wrong deliverables, and significant variances of schedule, budget, and quality. The goal of project management is predictable project performance, which depends on a project plan and project execution of the approved project plan.

15.14 Step 14: Congratulations

Congratulate yourself and your team for a project well planned and executed. These skills are useful for your personal projects and highly valued in the workplace. Update your resume, LinkedIn profile, and other summaries of your skill set. If you work for a company with personal development plans, be sure to advise the appropriate people of these new skill sets. If you enjoy this type of work, consider taking courses to enhance your skills in the project management discipline. You might also participate with professional associations such as Project Management Institute (www.pmi.org) and consider professional certifications.

Whether you choose to deploy these new skills in personal or workplace projects, your journey into project management has begun and will serve you and your projects well for many years.

Glossary

Acceptance Criteria: The conditions that must be satisfied for the key stakeholders to consider the project acceptable. This is likely less stringent than the success criteria. For example, if the project manager is running in a marathon race, he considers a completion time of less than 3 hours a success; he considers a completion time of less than 4 hours acceptable.

Analogous Estimating: An analogous estimate is a comparison to another similar project to develop an estimate in time or money. The project manager increases or decreases the estimate based on whether the comparison project is larger or smaller than the current project. It is a form of expert judgment and generally used for estimates early in the project life cycle, such as the project charter, when precision is less significant. Compare with parametric estimating and bottom-up estimating.

Applied Time: Sometimes called effort or work, applied time is the amount of time spent working on the task. This contrasts with elapsed time, which is the amount of time between the start and completion of a task.

Assumption: An assumption is something the project manager considers to be true for planning purposes. An assumption either proves to be true, in which case it is a fact. Or, it proves to be false and is a risk.

Bottom-Up Estimating: A bottom-up estimate is based on the project tasks associated with the work packages. It is precise but takes time to develop. The project team develops the bottom-up estimates; while they estimate the work, they do not have this time available to do the work of the project. Bottom-up estimating is appropriate at the end of the planning process when final project approval requires precise estimates. Compare with analogous estimating and parametric estimating.

Brainstorming: Teams use brainstorming to generate ideas. In response to a question or problem statement, participants share their ideas. The meeting facilitator records the ideas using the original wording. Other participants add to the idea or suggest new ideas. There is no comment or judgment on the original ideas; the ideas are captured as stated. After the brainstorming session, ideas may be consolidated to eliminate duplicates.

Budget: The budget is the project spending plan. The actual spending, called cost, is compared to the budget to determine positive and negative variances. Funding is the amount of money available. For the deck project, the cost estimated in the charter is \$3000; the project budget is \$2925 for materials. Actual costs are not specified, but the change request (Figure 14.1) proposes a cost increase of \$200.

Charter: The charter is a contract between the project sponsor and project manager that formalizes their relationship and project funding. Information in the charter may also be used to compare projects and prioritize funding across projects.

Constraint: A constraint limits options; for example, the deck project has a constraint that it must be at least 5 feet from abutting property.

Contingency Reserves: Funds set aside to cover the cost of the fallback plan.

Critical Path: This is the shortest amount of time in which all the work of the project can be completed. If any task on the critical path is completed late, then the end date of the project will slip unless the project manager takes action to compress subsequent tasks on the critical path.

Decomposition: The process of breaking a large item or concept into component parts. For example, the project manager decomposes the project core into scope, quality, schedule, and budget. The Community Gardens project may be decomposed into individual gardens, shared community garden, soup kitchen, farmers market, 4H competition, and cookbook.

Deliverable: This is the end result of a project; it may be a product, service, or result. It is frequently described in the noun–verb format to convey the idea of completion, such as deck completed or produce delivered to soup kitchen.

Duration: Duration is the amount of time from the start of a task until its completion, including nonwork time. For example, if the team member works an 8-hour day, an 8-hour task has a duration of 1 day. If the team member only works half-time (4 hours per day) on the task, that same task has a duration of 2 days. The 8 hours required for the task is effort, sometimes called applied time.

Effort: This is the time required to complete a task. It is sometimes called applied time.

Elapsed Time: Elapsed time is the amount of time between the start and completion of a task; it generally includes weekends, holidays, and other nonworking time.

Fallback Plan: Actions that the team takes if a risk occurs and the primary response doesn't work.

Impact: Generally used with risk, impact is the consequence of a risk; it is what happens as a result of a risk occurrence. For example, if the neighbors object to the deck (risk), the impact might be the failure to obtain the building permit. Contrast this with probability.

Metric: A measurement that allows the project manager to determine if an objective is met. For example, sponsor signature on the project charter is the metric for charter approval.

Milestone: The completion of a significant event within the project such as the approval of the building permit for the deck project or the initial harvest for the Community Gardens project.

Opportunity: An opportunity is a positive risk. Taking advantage of opportunities, the project team is able to exceed the project objectives. Contrast with threat (negative risk).

Parametric Estimating: A parametric estimate is based on a metric, such as square feet of garden that can be planted in an hour. Once the project manager determines the number of square feet of garden to be planted, he calculates the time required. Parametric guidelines are often publicly available on web sites. Compare with analogous and bottom-up estimating techniques.

Probability: Generally used with risk, probability is the likelihood that a risk will occur. Contrast this with impact.

Process: A specified set of steps intended to generate a service, product, or attain a goal.

Product: The product is the reason for the project; it takes the form of a thing, service, or result. The deck is the product of the deck project; vegetable produce is the product of the Community Gardens project.

Project: A project has a beginning and an end; it consumes a specified amount of time and creates a unique product, service, or result.

Project Boundaries: Documented in the project charter, project boundaries identify the limits of the project. For example, the boundary of the deck project is the deck itself; railing, lighting, and stairs are beyond the boundaries. See also project exclusions.

Project Context: The project context is the environment in which the project occurs. The project manager determines if the project context supports the project or creates challenges to project success. He manages the project to optimize the probability of success within the existing context.

Project Core: The project core includes scope, schedule, budget, and quality. Project scope is unique to each project, and schedule, budget, and quality are specific to the project scope.

Project Exclusions: Documented in the project scope statement, exclusions are specific items that are outside the scope of the project. Project exclusions are often listed to define the project boundaries.

Project Life Cycle: The project life cycle is the series of phases that all projects experience including *approval and start up*, *planning, executing, tracking and fixing,* and *finishing up.* Phases often include specific activities and skill sets; project managers monitor the completion of phase activities and manage the progression to the next phase based on the completion of prior phase activities.

Project Manager: The project manager is responsible for project success; he directs project planning and management. Sometimes, the project manager is also responsible for the product of the project.

Project Plan: The project plan includes all documentation developed to manage and execute the project. Table 13.2 lists the components of a typical project plan.

Quality: Quality refers to meeting stakeholders' expectations and requirements. There is both project quality and product quality.

Quality Plan: The quality plan documents the metrics to assess whether the project or product is a quality deliverable.

Residual Risk: These are risks that remain after the team implements the risk response plan.

Resources: The project manager uses resources to develop and implement the project plan. Resources include labor, materials, and services.

Responsibility Assignment Matrix (RAM): A RAM is a matrix that specifies the relationship between stakeholders and either project deliverables or documents. Examples of the relationship include responsible, consult, sign off, and inform.

Risk Response: A risk response is the strategy for dealing with a risk. Risk responses for project opportunities include exploit, enhance, share, or accept. Risk responses for project threats include avoid, mitigate, transfer, or accept.

Risk Score: The project manager uses the risk score to prioritize the positive and negative risks relative to each other. The risk score is the probability multiplied by the impact.

Risks: A risk is a future event; it may have a positive impact on the project (an opportunity) or a negative impact on the project (a threat).

Schedule: The project schedule lists the start and finish dates for the project overall and for all tasks within the project.

Schedule Network Diagram: The schedule network diagram illustrates the project tasks in sequence showing predecessor and successor tasks.

Scope: Project scope identifies the project and product deliverables. This is the "what" of the project (as opposed to "how").

Scope Statement: The scope statement lists the minimum information necessary to validate the go-no-go decision made in the project charter; it is the basis for the schedule, budget, and quality for the project.

Secondary Risk: A secondary risk may occur as a consequence of implementing the risk response plan. The project manager considers the possible secondary risks when evaluating a risk response plan.

Sponsor: The project sponsor is generally the project champion and responsible for funding the project with money and/or resources (labor, materials, and services). The project sponsor obtains necessary project support from other stakeholders.

Stakeholder: Stakeholders are involved with or impacted by the project. They have a *stake* in the project. Key stakeholders have influence over the decision or activity under consideration.

Stakeholder Engagement Matrix: The stakeholder engagement matrix identifies the stakeholders' current project perspective and the perspective preferred by the project manager. The project manager identifies a plan to move stakeholders from their current level of engagement to the preferred level of engagement.

Stakeholder Matrix: The stakeholder matrix lists the stakeholders and their scores for power and interest. It also lists what the stakeholder needs from the project and what the project needs from the stakeholder. The matrix clarifies the stakeholder WIIFM and the project needs.

Success Criteria: There are the conditions that must be met for the key stakeholders to consider the project a success. This is likely more stringent than the acceptance criteria, which are the conditions that must be met for the sponsor to pay for the project and the customer to accept it. If the project manager is running in a marathon race, he considers a completion time of less than 3 hours a success; he considers a completion time of less than 4 hours acceptable.

Task: Tasks are the work necessary to create a work package deliverable. While the deliverables are the "what" of a project, the tasks represent the "how" of a project.

Teaming: This is the process of getting the stakeholders to work collaboratively as a team and for the entire team to generate value that is greater than the aggregate value of the individual team members.

Template: A preset format for a document; it ensures that document content is consistent across projects, and that document content does not need to be recreated for each project.

Threat: A threat is a negative risk. Minimizing threats, the project team is able to achieve the project objectives. Contrast with opportunity (positive risk).

Toolkit: The project toolkit contains documents, standards, tactics, and processes that provide a framework for managing the project. They are repeatable and provide the project manager with a familiar starter kit for each project.

Variance: Variance is the difference between the expected and the actual value of project parameters such as schedule and budget. Variances may be positive or negative. The project manager establishes allowable variances and corrective actions appropriate to variances that exceed the allowable amount.

Virtual Team: A virtual team does not meet face-to-face. Team members may be geographically remote from each other or may simply work at different facilities or from their homes. The lack of face-to-face contact poses challenges to project communication and teaming.

Urgency (risk): This is the time criticality of a risk occurrence. It is sometimes used with probability and impact to create the risk score. The risk of a flood is more significant (*urgent*) if the river is currently cresting its banks than if the river overflow is several hours away.

Work Breakdown Structure (WBS): The WBS is a decomposition of the project deliverables to work packages. The WBS is generally created in a graphic format and may be documented in outline form.

Work Package: This is the lowest level of the WBS. Work packages are assigned to an appropriate team member to identify the tasks necessary to develop the work package deliverable.

Appendix A: Community Gardens Project

The Community Gardens (CG) is a collaborative gardening project on town property. Part of the garden is individual plots; part of the garden is a community plot and must be tended by all members of the Community Gardens project, most of whom also have individual plots. The produce from the Community Gardens serves the local soup kitchen, and the Community Gardens project sells produce at the farmers market on the town commons throughout the gardening season. The Community Gardens project enters prize winning vegetables in the local 4H competition. Many towns compete in the 4H competition, with one of two towns generally winning over other towns and contest entries. The two towns compete aggressively for the generous cash award for winning entries and for "bragging rights" throughout the year.

The towns have a long history of taking care of the less fortunate in the community; the soup kitchens rely on the fresh produce. The revenue from selling produce at the farmers market, along with profits from the Community Gardens cookbook, defrays the cost of the Community Gardens supplies and provides cash contributions to the town soup kitchen.

Each town is willing to contribute town resources for the pre- and post-season preparations; however, since town resources are used, the required resources must be included in the town budget. Otherwise, the town cannot legally allocate town resources for the pre- and post-season Community Gardens preparation.

A.1 Community Gardens Project

The project schedule starts immediately after the prior year's lessons-learned meeting and season-end barbeque to celebrate the team's hard work and success. Typically, the project manager holds a team meeting in late September or October. The project manager submits the budget request to the town in time for inclusion in the town budget and its approval process. Things are quiet throughout the winter months; planning for spring activities begins in February by developing plans and schedules. The season officially begins in March, and the project manager solicits gardeners for the individual gardens and also for the Community Gardens. At the kick-off meeting in mid-April, the team gathers to discuss what to plant and shared responsibilities during the season.

In May, the town prepares the gardens, turning, grading, and preparing the soil, setting up hoses for water, turning on the water, and putting up fences to keep out little critters and other unwelcome visitors. The project manager collects the requirements for seeds and plants from the individual gardeners and for the community plot. Individual gardeners may purchase their materials and supplies on their own or may pool their requirements with those of other individual gardeners and requirements of the community plot for reduced costs. The Community Gardens often benefit from reduced pricing when it purchases in large quantities; some vendors also provide significant discounts in exchange for advertising spots in the Community Gardens cookbook produced at the end of the season and sold at town events throughout the year. If individual gardeners choose to share in the bulk purchase, the Community Gardens team collects money in May for the materials and supplies.

The gardeners generally start planting in late May or early June after all risk of frost is gone; some folks plant hardier produce earlier. The Community Gardens have produce in mid-June and start weekly deliveries to the soup kitchen on Mondays and weekly participation in the farmers market on the town square on Thursdays. Individual gardeners transport the produce to the soup kitchen and staff the booth at the farmers market.

The 4H competition is an exciting event. Several towns have been in head-to-head competition for many years. First place in the 4H competition brings bragging rights to the winner and also a significant cash prize that is used along with the revenue from the farmers market and cookbook sales to defray expenses. Any additional money is shared among the soup kitchen and the individual gardeners. Generally, the individual gardeners elect to contribute their individual shares to the soup kitchen.

At the end of the season, the team gathers for a lessons-learned meeting and season-end celebration. Everyone tastes entries from all recipe candidates for the cookbook (which must use produce from the gardens) at the year-end celebration; only approved recipes are included in the cookbook.

A.2 The Community Gardens Project: Two Scenarios

Two towns, Legacy and Planful, are generally the top contenders for the 4H first place prize. The Community Gardens project managers for each town are a bit different in approach. Larry manages Legume Legacy Community Gardens. Larry has been managing the Community Gardens for 10+ years. He is 56 years old, semi-retired with lots of time on his hands. His style is "old boy network," seat of the pants, and he is used to a "command and control" structure.

George leads Planful Community Gardens for his second year. He is in his 30s and is a project manager at his full-time job. He is very focused and structured. His style is collaborative/consensus.

George and his Planful Gardens project team follow a project process and develop project documents that, in the aggregate, form a project plan. The Community Gardens project documents in the appendices that follow are typical of the documents created by George and the Planful Community Gardens team. Appendix R contains project execution based on the project plan for George and no project plan for Larry and his team.

Appendix B: Project Charter

George meets with the town manager as sponsor of the Planful Community Gardens. Since the project charter is the sponsor's statement of expectations and constraints of the Community Gardens project, George encourages the town manager to take an active role in the creation of the project charter. Unfortunately, the town manager, like sponsors of many projects, is reluctant to take an active role and delegates the creation of the project charter to George.

Working with the sponsor and other key stakeholders, George develops the project charter for the Community Gardens as shown in Figure B.1.

Project Charter

Project Name: Community Gardens

Project Purpose and Requirements

Purpose

- Support community less fortunate with food via soup kitchen
- Enhance town reputation by winning the 4H competition
- Enable town residents to have inexpensive, healthy produce

Requirements

- Land and land preparation contributed by the town
- Water supplied by town

Objectives and Success Criteria

- Contribute a minimum of 340 pounds of produce to the local soup kitchen each week for 12–14 weeks from mid-June through mid-September (from Community Gardens)
- Generate a minimum of $3600 from sale of produce at local farmers market (from Community Gardens) during the season
- Win first place for produce at 4H competition
- Individual gardeners generate an aggregate of 0.5 pounds of produce for each square foot of ground planted

Description and Boundaries

- This is a community garden, shared between individual plots and a community plot. The primary focus is produce, although individual gardeners may use their land for flowers. The garden includes fences, water, hoses, and plots that are marked. There is no requirement to use organic gardening techniques, but nontoxic and environmentally friendly gardening methods are encouraged
- Out of scope—the Community Gardens does not include lighting, parking, traffic control, and pest/varmint control

Key Risks

- There is a risk that there will not be enough interest and not enough gardeners to work on the individual plots and support the community section
- There is a risk that the harvest will not be good and there will not be enough produce to support the soup kitchen and the farmers market
- There is a risk that there will not be enough volunteers to staff the weekly farmers market

Assumptions

- The weather will be no better or worse than average (as specified by the Farmers' Almanac, a trusted source of weather forecasts)
- The farmers market in the town will continue on a weekly basis
- The soup kitchen will continue to serve the town's less fortunate residents

Figure B.1 Project charter for Community Gardens project. *(Continued)*

Constraints

- The amount of land is fixed; the project manager can increase/decrease the number of individual plots and/or alter the allocation of land between the Community Gardens and the individual plots, but there is no more land available
- There is no lighting
- The project uses the municipal accounting system so the project is subject to the town's rules and limited to vendors approved by the town

Milestones

- Season kickoff—mid-April
- Money available to purchase materials/supplies—early May
- Planting started—late May
- Harvest begins—mid-June
- 4H competition—late August
- Lessons learned—late September or early October

Budget

Costs

- Estimate cost of plants for Community Gardens at $2000
- Estimate cost of services—$800 for town services

Revenue

- Farmers market revenue = $3600
- Community Gardens cookbook = $1000
- Prize from 4H competition = $500

Stakeholders

- Project manager
- Individual gardeners
- Manager of soup kitchen
- Manager of farmers market
- Organizer of 4H competition
- Town manager
- Vendors

Acceptance Criteria

- Soup kitchen manager signs off on whether contributions met plan (quality and quantity)
- Project manager signs off on success of farmers market (revenue, staffing)
- Percent of returning gardeners (60% returning) is another indicator of success

Approvals and Date

Project Manager: either George for Planful Gardens or Larry for Legume Legacy
Sponsor: town manager

Figure B.1 (Continued) Project charter for Community Gardens project.

Appendix C: Project Scope Statement

George has much information for the project scope statement from last year's project. He has the original project scope statement and information from the lessons-learned meeting at the end of last year's project. He drafts a project scope statement based on this historical information and any changes anticipated for this year's Community Gardens project. He consults with key stakeholders to obtain current information and shares the completed document with the town manager. Once the scope statement is final, key stakeholders sign and date the document indicating their approval.

Figure C.1 contains the Community Gardens project scope statement.

Scope Statement

Project Name: Community Gardens

Scope Description

- A football field-sized garden plot (360 × 120 feet) shared by Community Gardens and individual plots. The land should be divided equally between the Community Gardens and individual plots (sized 25 × 25 feet), but the size of the Community Gardens and individual plots may be changed based on demand. However, the total amount of available land is fixed.
- Fencing will be 5 feet high and will go 1 foot below the surface to discourage animals from burrowing under the fence. There will be a gate at both the north and south sides of the garden.
- The town will provide and connect faucets to the town water supply; gardeners will need to provide sufficient length of hose to water the gardens.
- Mulch will be available from the town dump. Gardeners will need to pick this up from the dump unless arrangements can be made to use a town truck to deliver the mulch in bulk.

Acceptance Criteria

- Produce will be inspected for quality before delivery to the soup kitchen and the farmers market. The role of quality inspector will rotate among the individual gardeners. Final acceptance is by the soup kitchen manager and the customers at the farmers market. Acceptance includes purchasing produce from the farmers market, and acceptance of produce by the soup kitchen manager.
- Acceptance of town pre-season preparation is based on the action of the individual gardeners. Complaints and/or failure to plant the garden plots are indications of non-acceptance of preparation by the town.
- Entry of produce for the 4H competition; final acceptance will be based on winning the first prize.

Deliverables

- Produce from individual and Community Gardens (distributed to soup kitchen and farmers market)
- First prize in 4H competition
- Community Gardens cookbook
- Project documentation

Figure C.1 Project scope statement for Community Gardens project. *(Continued)*

Exclusions

- No lighting provided
- No additional parking—only existing dirt-based parking available
- No additional traffic control
- Town will not provide pest and varmint control; this is the responsibility of individual gardeners
- No trash pickup; debris must be taken to town dump

Constraints

- Maximum amount of land is limited
- Only town-approved vendors may be used
- Plants for Community Gardens must be purchased from town-approved vendor
- Produce may only be sold through the weekly town farmers market
- Produce can only be distributed to town-approved soup kitchen
- The project must use the town accounting system

Assumptions

- The town has maintenance staff available to prepare the land prior to the start of the planting season in mid-May
- Quality plants will be available
- Weather will be no better/worse than normal; gardeners will be able to compensate for adverse conditions through aggressive watering, fertilizing, pest control, or other standard gardening approaches

Approvals and Date:

George as Project Manager: ____________________________ **Date:** ________
Town Manager as Project Sponsor: ______________________ **Date:** ________
Gardeners as Customer: ________________________________ **Date:** ________

Figure C.1 (Continued) Project scope statement for Community Gardens project.

Appendix D: Work Breakdown Structure

Working with his team, George creates a work breakdown structure (WBS) chart using sticky notes. He includes individual gardeners involved in last year's project and invites all interested gardeners to join the effort. Using the list of stakeholders from the charter, scope statement, and preliminary stakeholder matrix, George invites the soup kitchen manager and the farmers market manager to join the meeting or provide their ideas directly. He invites the abutters to the Community Gardens to gain their support by involving them in the project effort. He makes sure the other stakeholders are aware of the meeting so that they might participate if interested. He also invites the town manager (as the project sponsor) and discusses the WBS with the town manager to ensure that the project has strong sponsor support.

Once the project team completes the WBS in sticky note format, George takes a photograph of the finished document.

Figure D.1 illustrates the Community Gardens project WBS in sticky note format. This is often called a tree structure. Note that the components of the WBS follow the noun–verb construction discussed in the text. Deliverables 4.0 and 5.0 are further decomposed into smaller work packages; some top level deliverables are work packages, and other top level deliverables are decomposed into smaller work packages.

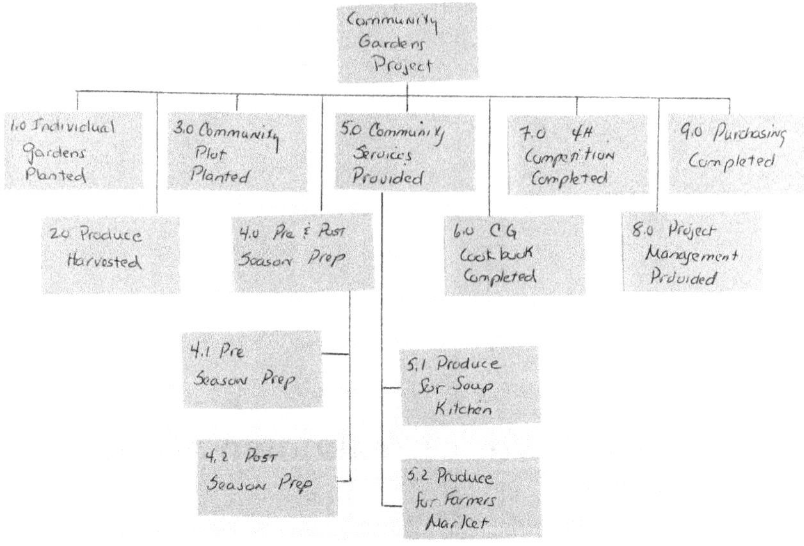

Figure D.1 WBS for Community Gardens project—Sticky note format.

George photographs the WBS in sticky note format for inclusion in the project plan. He may then type the information in the photograph; this is optional as the typed graphic contains the same information as the photograph. Figure D.2 illustrates the Community Gardens project WBS in a typed graphical format.

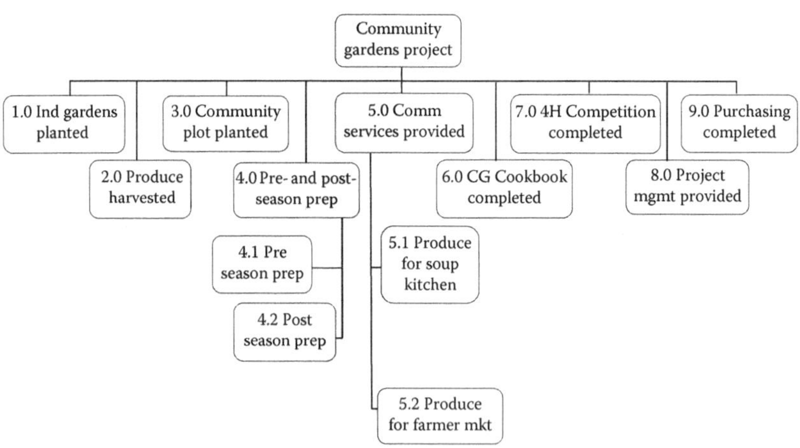

Figure D.2 WBS for Community Gardens project—Graphical format.

Table D.1 WBS for Community Gardens Project—Outline Format

COMMUNITY GARDENS PROJECT

- 1.0 Individual Gardens Planted
- 2.0 Produce Harvested
- 3.0 Community Plot Planted
- 4.0 Pre- and Post-Season Preparation
    - 4.1 Pre-Season Preparation
    - 4.2 Post-Season Preparation
- 5.0 Community Services Provided
    - 5.1 Produce for Soup Kitchen Delivered
    - 5.2 Produce for Farmers Market Sold
- 6.0 Community Garden Cookbook
- 7.0 4H Competition Completed
- 8.0 Project Management Provided
- 9.0 Purchasing Completed

George reorganizes the information into an outline format; this is sometimes called an indented hierarchy. It is exactly the same information but presents the information in an outline format rather than a graphical format. Stakeholders process information differently; the project manager offers the same information in different formats to ensure that he speaks in the language with which the stakeholder is most conversant to ensure understanding and support.

Table D.1 illustrates the Community Gardens WBS in an outline format.

Once these are complete, George shares them with key stakeholders and obtains their sign off and approval.

Appendix E: Work Breakdown Structure with Tasks

Having secured stakeholder approval for the Community Gardens WBS, George invites the project team to his home for a working session to develop the extended WBS. It is important that the team members who will do the work participate in the identification of the tasks associated with the work packages; refreshments, in the form of pizza, seem to promote attendance, hard work, and good results. George posts the approved WBS on the wall and leaves packets of sticky notes around the table. He introduces the WBS work packages and suggests that certain team members take ownership of work packages in their area of expertise. He reminds them of the verb–noun format appropriate to tasks and suggests that the team discuss and identify the tasks necessary to deliver these work packages.

The team members use sticky notes to capture and organize the tasks under the appropriate WBS work packages; the sticky note format allows flexibility in organizing and moving tasks around while discussing and creating the extended WBS. Once the team is satisfied that the extended WBS is complete, George photographs the extended WBS (Figure E.1).

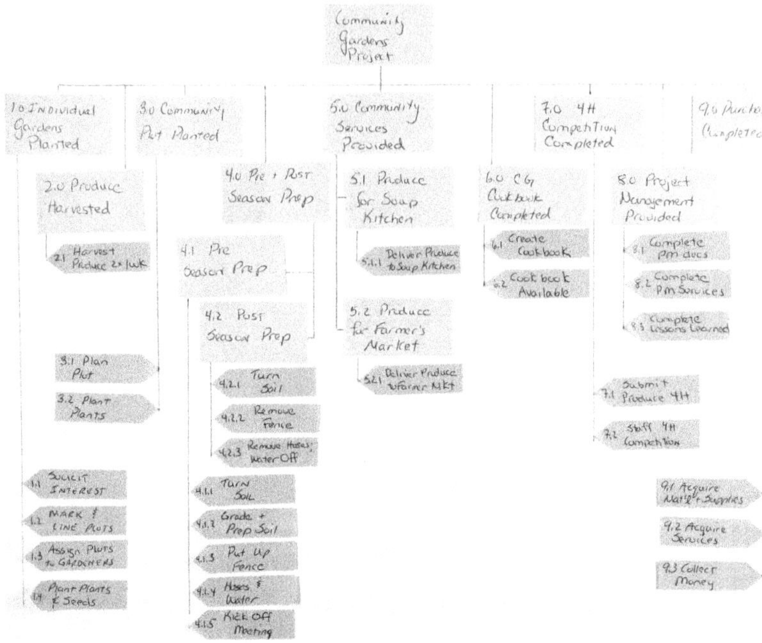

Figure E.1 Extended WBS for Community Gardens project—Sticky note format.

George may then create a typed graphical version of the photograph and also present the extended WBS as an indented hierarchy (Figure E.2 and Table E.1).

As with other project documents, George shares them with appropriate stakeholders and obtains approvals.

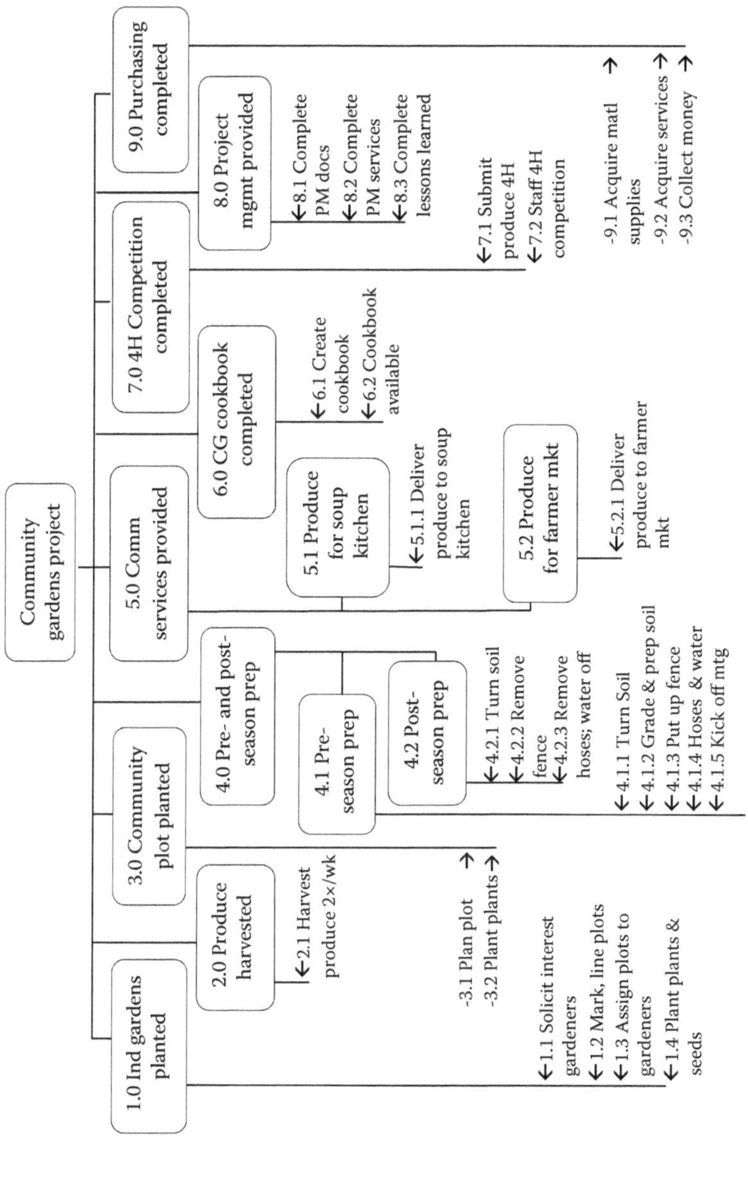

Figure E.2 Extended WBS for Community Gardens project—Graphical format.

Table E.1 Extended WBS for Community Gardens Project—Outline Format

COMMUNITY GARDENS PROJECT

- 1.0 Individual Gardens Planted
  - 1.1 Solicit Interest Gardeners
  - 1.2 Mark Line Plots
  - 1.3 Plots Assigned to Gardeners
  - 1.4 Plant Plants and Seeds
- 2.0 Produce Harvested
  - 2.1 Harvest Produce 2×/week
- 3.0 Community Plot Planted
  - 3.1 Plan Plot
  - 3.2 Plant Plants
- 4.0 Pre- and Post-Season Preparation
  - 4.1 Pre-Season Preparation
    - 4.1.1 Turn Soil
    - 4.1.2 Grade and Soil Preparation
    - 4.1.3 Put Up Fence
    - 4.1.4 Hoses and Water
    - 4.1.5 Kick Off Meeting
  - 4.2 Post-Season Preparation
    - 4.2.1 Turn Soil
    - 4.2.2 Remove Fence
    - 4.2.3 Remove Hoses and Water Off
- 5.0 Community Services Provided
  - 5.1 Produce for Soup Kitchen Delivered
    - 5.1.1 Deliver Produce to Soup Kitchen
  - 5.2 Produce for Farmers Market Sold
    - 5.2.1 Deliver Produce to Farmers Market
- 6.0 Community Garden Cookbook
  - 6.1 Create Cookbook
  - 6.2 Cookbook Available
- 7.0 4H Competition Completed
  - 7.1 Submit Produce 4H
  - 7.2 Staff 4H Competition
- 8.0 Project Management Provided
  - 8.1 Complete PM Documents
  - 8.2 Complete PM Services
  - 8.3 Complete Lessons Learned
- 9.0 Purchasing Completed
  - 9.1 Acquire Materials and Supplies
  - 9.2 Acquire Services
  - 9.3 Collect Money

Appendix F: Schedule Network Diagram

After the project stakeholders approve the Community Gardens extended WBS, George reconvenes his core team to sequence the project tasks. He repeats the informal, pizza-supported meeting; he posts the extended WBS on the wall and places sticky notes around the work spaces. The team writes each task description and number on a sticky note and begins the discussion of appropriate sequencing of the tasks. The tasks are the same as listed on the extended WBS but will be sequenced in predecessor–successor format rather than organized under the WBS work packages to which the tasks belong. George stays in the background, letting the team take the lead in the discussion. He participates as necessary to moderate the conversation and keep it on track and focused.

Once the team is satisfied with the schedule network diagram, George takes a photograph to preserve the work and shares it with other stakeholders (Figure F.1).

George may elect to type this information so it is easier to read and circulate to stakeholders for review and approval (Figure F.2).

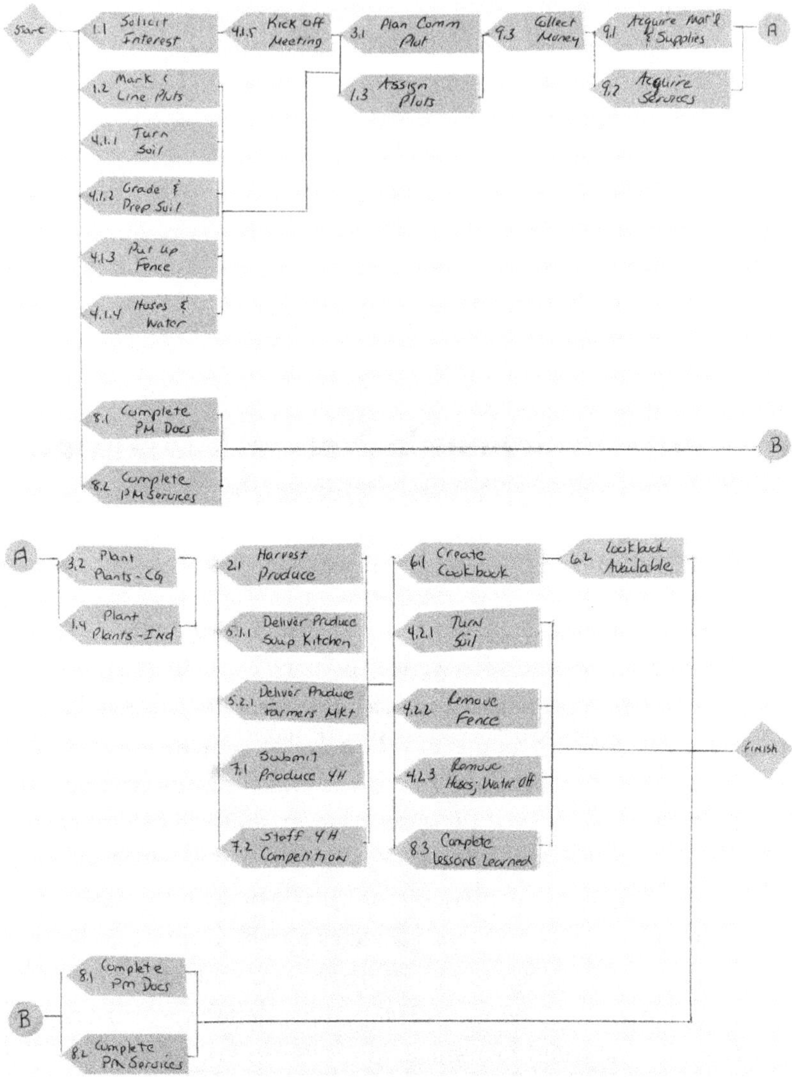

Figure F.1 Schedule network diagram for Community Gardens project—Sticky note format.

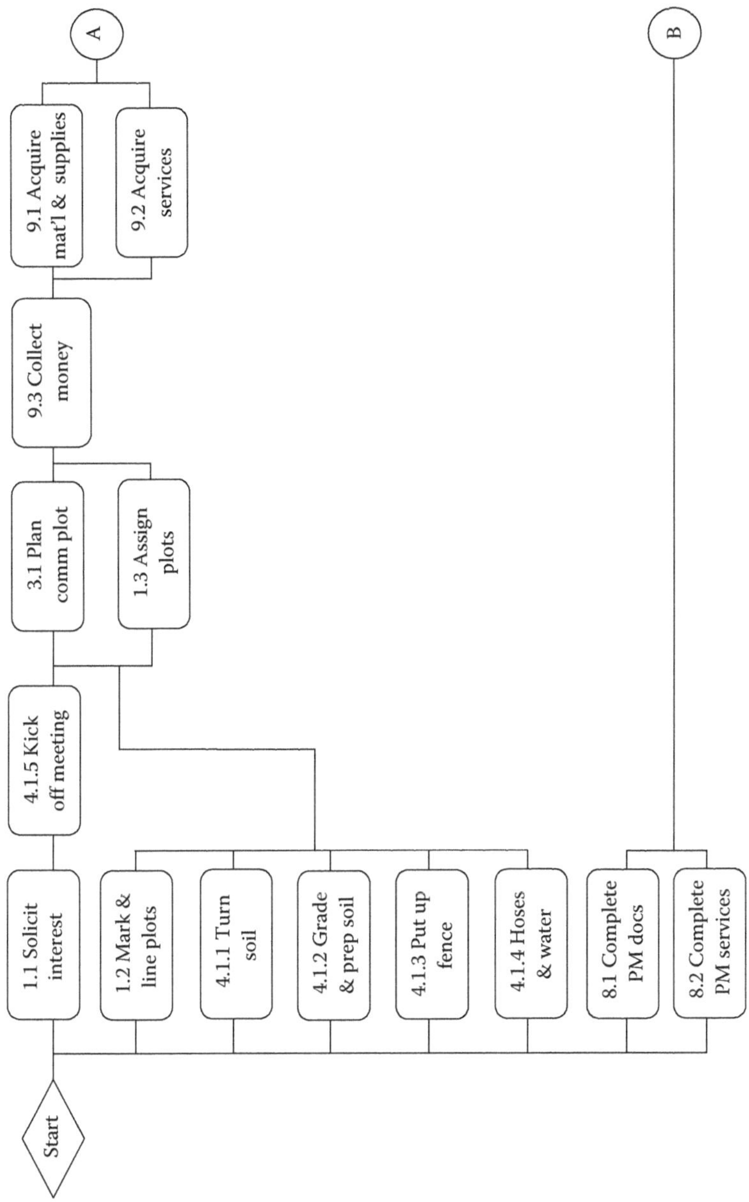

Figure F.2 Schedule network diagram for Community Gardens project—Graphical format.

(Continued)

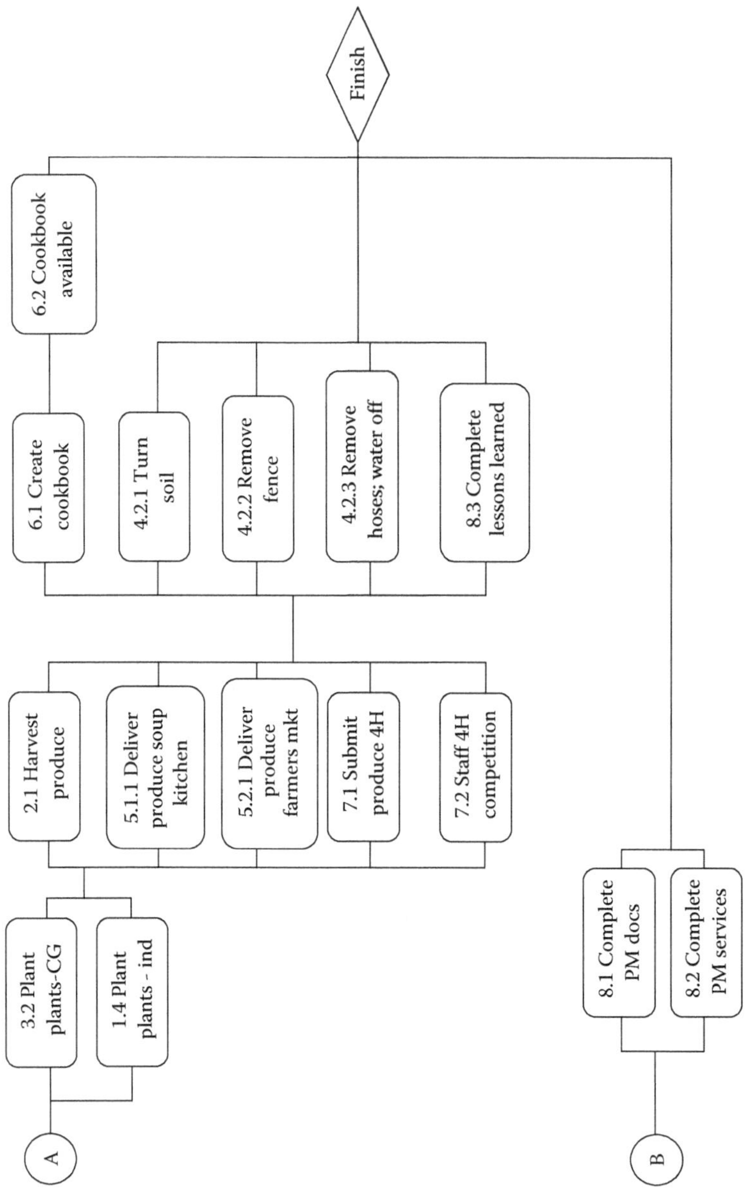

Figure F.2 (Continued) Schedule network diagram for Community Gardens project—Graphical format.

Appendix G: Project Schedule

Based on the sequenced schedule network diagram, George completes the project schedule. As with the deck example in the text, the project schedule includes, for each task, the applied time (effort), percent of time, duration, and resources, both number of resources and skills/ specialties, if appropriate.

G.1 Time Increments

For this project, the schedule tracks time by week, rather than by day (as is the case for the deck project). For longer duration projects, it is challenging to track work in small time increments; the rule of thumb is to increase the time increment to weeks, months, or quarters. For very long duration projects, the schedule may initially plan and track in weeks, months, or quarters; for the near term tasks, the team adjusts the plan to smaller time increments (e.g., from weekly to daily or monthly to weekly), and the team monitors progress against the smaller time increments.

If the team chooses weeks as the project's time increment, the schedule specifies how many work days in a week. For example, a week might include 4, 5, or 6 eight-hour days; the plan specifies how many days in a week and how many hours in a work day. If some weeks include holidays or other nonworking time that reduces the week's hours available for work, the schedule specifies this also. Figure G.1 includes a row that specifies the number of days available each week. For the Community Gardens project, the default project work week is 5 days, and there are 8 work hours available each workday.

As the team allocates tasks to each week, it must consider the work of each task. A 5-day week may include up to 5 days of duration for a single resource.

G.2 Assumptions

Figure G.1 illustrates the schedule for the Community Gardens project. The Community Gardens project planning includes a number of assumptions, as follows:

- There are 25 individual garden plots; the plan assumes that there will be 25 individual gardeners and they all share the work of the Community Gardens, in addition to working their own plots.
- The project manager assumes that it takes each gardener 8 hours to plant the individual garden; since the Community Gardens is the size of 25 individual gardens, the plan calculates 200 hours (25 plots × 8 hours per garden plot) to plant the Community Gardens.
- For task 2.1, Harvest Produce, the plan assumes that the team will need 15 hours each week to harvest for the soup kitchen and the same amount of time to harvest for the farmers market. This is 30 hours applied time (effort) each week for 16 weeks. If 5 individual gardeners work the weekly harvest for the soup kitchen and 5 gardeners work the weekly harvest for the farmers market, it will take 3 hours duration per harvest.
- The project manager estimates time for project management work as a percent of all other tasks; 10% is a reasonable estimate. For both project management tasks 8.1 and 8.2, he allocates more of the time to the first and last weeks of the project as specified in the budget (Figure H.1).
- During week 1 of the project, there are 6 tasks, each with a duration of 1 day. If the same resource is responsible for all 6 tasks, the team would not be able to complete the six

	Applied					4/14	4/21	4/28	5/5	5/12	5/19	5/26	6/2	6/9	6/16	6/23	6/30	7/7	7/14	7/21
Task	Time	% Time	Durat	Resource	Dur Days	5	5	5	5	5	5	5	5	5	5	5	5	5	5	5
					Cal Week	1	2	3	4	5	6	7	8	9	10	11	12	13	14	15
1.1 Solicit Interest-Gardeners	6 hr		1 d	team		x														
4.1.5 Kick Off Meeting	2 hr		1 d	PM			x													
1.2 Mark & Line Plots	6 hr		1 d	town		x														
4.1.1 Turn Soil	6 hr		1 d	town		x														
4.1.2 Grade & Prep Soil	6 hr		1 d	town		x														
4.1.3 Put Up Fence	4 hr		1 d	town		x														
4.1.4 Hoses & Water	4 hr		1 d	town		x														
3.1 Plan Comm Plot	5 hr		1 d	team				x												
1.3 Assign Plots	2 hr		1 d	team				x												
9.3 Collect Money	5 hr		1 d	PM					x											
9.1 Acquire Mat'l	8 hr		1 d	team						x										
9.2 Acquire Services	3 hr		1 d	Gardener						x										
3.2 Plant Plants (CG)	200 hr	50%	2 d	25 Gdnrs							x									
1.4 Plant Plants (Ind)	200 hr	50%	2 d	25 Gdnrs							x									

Figure G.1 Schedule for the Community Gardens project.

(*Continued*)

	Schedule					4/14	4/21	4/28	5/5	5/12	5/19	5/26	6/2	6/9	6/16	6/23	6/30	7/7	7/14	7/21
Task	App Tim	% Time	Durat	Resource	Dur Days	5	5	5	5	5	5	5	5	5	5	5	5	5	5	5
					Cal Week	1	2	3	4	5	6	7	8	9	10	11	12	13	14	15
2.1 Harvest Produce 2x/wk 30 hr*16 wk	480 hr	35%	16 wk	5 Gdnrs											x	x	x	x	x	x
5.1.1 Deliver Produce Soup Kitchen 4 hr * 16 wk	64 hr		16 wk	2 Gdnrs											x	x	x	x	x	x
5.2.1 Deliver Produce Farmer Mkt 4 hr * 16 wk	64 hr		16 wk	2 Gdnrs											x	x	x	x	x	x
7.1 Submit Produce 4H																				
7.2 Staff 4H Competition																				
6.1 Create Cookbook																				
6.2 Cookbook Available																				
4.2.1 Turn Soil																				
4.2.2 Remove Fence																				
4.2.3 Remove Hoses/Water																				
8.1 Complete PM Docs*	36 hrs		28 wk	PM		x	x	x	x	x	x	x	x	x	x	x	x	x	x	x
8.2 Complete PM Svs†	60 hrs		28 wk	PM		x	x	x	x	x	x	x	x	x	x	x	x	x	x	x
8.3 Complete LL																				

Notes: * = Assumes 1st and last 2 weeks at 3 hours, 24 weeks at 1 hour; † = Assumes 1st and last 2 weeks at 3 hours, 24 weeks at 2 hours.

Figure G.1 (Continued) Schedule for the Community Gardens project.

(Continued)

	Schedule					7/28	8/4	8/11	8/18	8/25	9/1	9/8	9/15	9/22	9/29	10/6	10/13	10/20		
Task	App Tim	% Time	Durat	Resource	Dur Days	5	5	5	5	5	5	5	5	5	5	5	5	5		
					Cal Week	16	17	18	19	20	21	22	23	24	25	26	27	28		
2.1 Harvest Produce 2x/wk	480 hr	35%	16 wk	5 Gdnrs		x	x	x	x	x	x	x	x	x	x					
5.1.1 Deliver Produce Soup Kitchen	64 hr		16 wk	2 Gdnrs		x	x	x	x	x	x	x	x	x	x					
5.2.1 Deliver Produce Farmer Mkt	64 hr		16 wk	2 Gdnrs		x	x	x	x	x	x	x	x	x	x					
7.1 Submit Produce 4H	6 hr		1 d	PM						x										
7.2 Staff 4H Competition	8 hr		1 d	tea m						x										
6.1 Create Cookbook	10 hr		3 d	team													x			
6.2 Cookbook Available	0 hr			team														x		
4.2.1 Turn Soil	6 hr		1 d	town													x			
4.2.2 Remove Fence	3 hr		1 d	town													x			
4.2.3 Remove Hoses; Water off	3 hr		1 d	town													x			
8.1 Complete PM Docs	36 hrs		28 wk	PM		x	x		x	x	x	x	x	x	x	x	x	x		
8.2 Complete PM Svs	60 hrs		28 wk	PM		x	x	x	x	x	x	x	x	x	x	x	x	x		
8.3 Complete LL	4 hrs		1 d	PM													x			

Figure G.1 (Continued) Schedule for the Community Gardens project.

1-day duration tasks in 5 days. However, the town handles the pre- and post-season preparation, and a project team resource will manage task 1.1, Solicit Gardener Interest. The pre-season preparation tasks will likely be completed one task per day by the town maintenance department. Task 1.1 may be completed on any day during that first project week. The project manager also works 3 hours on project documents and 3 hours on project services during this first week.

- During week 2, the only activity (other than project management work) is the kick off meeting. The project manager could choose to begin the successor tasks during this week; he has decided to wait until week 3.
- During week 3, there are only a couple of tasks, assigning plots to individual gardeners and planning the Community Gardens. There is unallocated time during this week, and the team could begin tasks 9.3, 9.1, and 9.2 during this week. The project manager chooses not to do this. The Community Gardens are in the northeast, where climate suggests that the team not plant outdoors until late May. If stakeholders prefer, they might start tasks 9.3, 9.1, and 9.2 sooner so that planting may start earlier.
- Task 3.2, Plant Plants (CG), assumes that all 25 gardeners will work 8 hours each ($25 \times 8 = 200$ hours) so the team can complete the planting in a week. Similarly, task 1.4, Plant Plants (Ind), assumes that all 25 gardeners will spend 8 hours each planting their own garden. Since this is 16 hours per gardener, both tasks are completed in the same week (week 6). Note that the gardeners are only working half time so these tasks have a 2 day duration.
- There is a gap of 3 weeks between the time the gardeners plant and produce is available for harvesting. Most likely, the gardeners will harvest from plants and not from seedlings.
- In week 10, the team begins to harvest produce and deliver to the soup kitchen and farmers market. Although task 2.1, Harvest Produce, precedes tasks 5.1.1, Deliver to Soup Kitchen, and 5.2.1, Deliver to Farmers Market, the team plans to harvest early in the week for deliveries to the soup kitchen,

and then again later in the week for deliveries to the farmers market. These tasks repeat for the 16 weeks of harvesting.

- The schedule network diagram shows tasks 5.1.1, 5.2.1, 7.1, and 7.2, occurring in parallel (at the same time). The schedule for tasks related to the 4H competition (tasks 7.1 and 7.2) is driven by the dates for the 4H competition and availability of produce.
- The post-season preparation (work package 4.2) must wait until the gardeners complete the harvesting; tasks 4.2.1, 4.2.2, and 4.2.3 follow the harvest and produce deliveries. Along with the post-season preparation, the team creates the cookbook and conducts the lessons learned. Theoretically, the team might schedule the post-season preparation tasks for the week immediately following the last weekly harvest; they elect to delay the post-season tasks for a week in case there is additional produce to be harvested. This one week delay doesn't affect the project end date.
- The team schedules task 8.3, Lessons Learned, a few weeks before project end. This allows the project manager to follow up on action items and close out the project in an orderly manner.

Appendix H: Project Budget

George creates the project budget from the approved project schedule. He takes the chart for the project schedule and replaces the "x," designating when a particular task occurs with the cost. A task may have several "x"s indicating that the task occurs over a period of several weeks. Cost may be designated in dollars or in time. For the Community Gardens, all labor is volunteer time, so George calculates labor costs in hours, with the hours corresponding to applied time. He spreads the applied time across the weeks during which the team works on the task. Material costs are actual out-of-pocket costs, and the project funds these tasks. George lists these in dollars.

The project budget for the Community Gardens project (Figure H.1) might look as follows:

- Labor Hours—there is a total of 1204 hours, with 96 of these hours for project management tasks 8.1 and 8.2. Project management tasks are about 9% of other project costs; this is a reasonable investment to increase the probability of project success. Think of the project management costs as an insurance policy. The sponsor is willing to spend up to 10% of the project costs to optimize the probability that the remaining costs (90%) are used effectively. These labor costs include the resources of the project manager, the team, the gardeners (a subset of the team), and town resources.

- Materials/Supplies—there is a total of $6000 for materials. George allocates $5000 for plants, half for the Community Gardens and half for the individual gardens. He allocates $1000 for services, which might include the value of the resources contributed by the town for pre- and post-season preparation. He may also elect to use town vehicles for transporting produce to the soup kitchen and also to the farmers market. In the past, team volunteers have used their own vehicle, but there is concern that there may be liability involved in using personal vehicles. Town or rental vehicles mitigate this risk.

He further decomposes these costs into weekly or monthly amounts.

- $6000 of material costs occur during week 5 of the project (tasks 9.1 and 9.2). George collects $2500 from the individual gardeners for the plants for the individual gardens (task 9.3); the project needs funding for the $2500 for plants for the Community Gardens and $1000 for services. Ideally, he funds this from prior year's monies (sales of produce and Community Gardens Cookbook) held back to cover these expenses. Otherwise, he asks the vendor(s) to accept delayed payment—he'd like to pay once he begins to get revenue from the farmers market.
- There is a significant labor cost during week 6 when the gardeners plant the individual and Community Gardens. This would be a significant funding concern if the labor were not volunteer labor.
- Weeks 10 through 25 are fairly level as the team harvests the produce, delivers it to the soup kitchen, and sells it at the farmers market. Week 20 also includes labor costs for the 4H competition.

The costs in the project charter include a budget of $2000 for plants (specific to the Community Gardens and excluding the cost of plants for the individual gardens) and $800 for services. The current estimates are just over 25% higher at $3500 for materials and supplies.

	Applied					4/14	4/21	4/28	5/5	5/12	5/19	5/26	6/2	6/9	6/16	6/23	6/30	7/7	7/14	7/21
Task	Time	% Time	Durat	Resource	Dur Days	5	5	5	5	5	5	5	5	5	5	5	5	5	5	5
					Cal Week	1	2	3	4	5	6	7	8	9	10	11	12	13	14	15
1.1 Solicit Interest-Gardeners	6 hr		1 d	team		6														
4.1.5 Kick Off Meeting	2 hr		1 d	PM			2													
1.2 Mark & Line Plots	6 hr		1 d	town		6														
4.1.1 Turn Soil	6 hr		1 d	town		6														
4.1.2 Grade & Prep Soil	6 hr		1 d	town		6														
4.1.3 Put Up Fence	4 hr		1 d	town		4														
4.1.4 Hoses & Water	4 hr		1 d	town		4														
3.1 Plan Comm Plot	5 hr		1 d	team				5												
1.3 Assign Plots	2 hr		1 d	team				2												
9.3 Collect Money	5 hr		1 d	PM					5											
9.1 Acquire Mat'l	8 hr		1 d	team						8										
				mat'l						5000										
9.2 Acquire Services	3 hr		1 d	gardener						3										
				mat'l						1000										
3.2 Plant Plants (CG)	200 hr	50%	2 d	25 gdnrs							200									
1.4 Plant Plants (Ind)	200 hr	50%	2 d	25 gdnrs							200									

Figure H.1 Budget for Community Gardens project. *(Continued)*

						4/14	4/21	4/28	5/5	5/12	5/19	5/26	6/2	6/9	6/16	6/23	6/30	7/7	7/14	7/21
Task	App Tim	% Time	Durat	Resource	Dur Days	5	5	5	5	5	5	5	5	5	5	5	5	5	5	5
					Cal Week	1	2	3	4	5	6	7	8	9	10	11	12	13	14	15
2.1 Harvest Produce 2x/wk 30 hr*16 wk	480 hr	35%	16 wk	5 gdnrs											30	30	30	30	30	30
5.1.1 Deliver Produce Soup Kitchen 4 hr * 16 wk	64 hr		16 wk	2 gdnrs											4	4	4	4	4	4
5.2.1 Deliver Produce Farmer Mkt 4 hr * 16 wk	64 hr		16 wk	2 gdnrs											4	4	4	4	4	4
7.1 Submit Produce 4H																				
7.2 Staff 4H Competition																				
6.1 Create Cookbook																				
6.2 Cookbook Available																				
4.2.1 Turn soil																				
4.2.2 Remove Fence																				
4.2.3 Remove Hoses/Water																				
8.1 Complete PM Docs*	36 hrs		28 wk	PM		3	3	1	1	1	1	1	1	1	1	1	1	1	1	1
8.2 Complete PM Svs†	60 hrs		28 wk	PM		3	3	2	2	2	2	2	2	2	2	2	2	2	2	2
8.3 Complete LL																				

Notes: * = Assumes 1st and last 2 weeks at 3 hours, 24 weeks at 1 hour; † = Assumes 1st and last 2 weeks at 3 hours, 24 weeks at 2 hours.

Figure H.1 (Continued) Budget for Community Gardens project. *(Continued)*

						7/28	8/4	8/11	8/18	8/25	9/1	9/8	9/15	9/22	9/29	10/6	10/13	10/20
Task	App Tim	% Time	Durat	Resource	Dur Days	5	5	5	5	5	5	5	5	5	5	5	5	5
					Cal Week	16	17	18	19	20	21	22	23	24	25	26	27	28
2.1 Harvest Produce 2x/wk	480 hr	35%	16 wk	5 gdnrs		30	30	30	30	30	30	30	30	30	30			
5.1.1 Deliver Produce Soup Kitchen	64 hr		16 wk	2 gdnrs		4	4	4	4	4	4	4	4	4	4			
5.2.1 Deliver Produce Farmer Mkt	64 hr		16 wk	2 gdnrs		4	4	4	4	4	4	4	4	4	4			
7.1 Submit Produce 4H	6 hr		1 d	PM						6								
7.2 Staff 4H Competition	8 hr		1 d	team						8								
6.1 Create Cookbook	10 hr		3 d	team													10	
6.2 Cookbook Available	0 hr			team														0
4.2.1 Turn Soil	6 hr		1 d	town													6	
4.2.2 Remove Fence	3 hr		1 d	town													3	
4.2.3 Remove Hoses; Water off	3 hr		1 d	town													3	
8.1 Complete PM Docs	36 hrs		28 wk	PM		1	1	1	1	1	1	1	1	1	1	1	3	3
8.2 Complete PM Svs	60 hrs		28 wk	PM		2	2	2	2	2	2	2	2	2	2	2	3	3
8.3 Complete LL	4 hrs		1 d	PM													4	

Figure H.1 (Continued) Budget for Community Gardens project.

(This is based on $5000 for plants plus $1000 for services minus $2500 contributed to the cost of plants by individual gardeners.) Balanced against the income projected in the project charter of $5100, this unfavorable variance may be acceptable. George seeks stakeholder approval to proceed with this year's Community Gardens based on these budget numbers. He also makes a note to revisit this variance in the lessons learned at the end of the project. He thinks that the original estimate in the charter was based on last year's costs, which were subsidized by vendor contribution of plants.

Appendix I: Project Quality Plan

George convenes a meeting of the project sponsors and customers, including the soup kitchen manager, a representative of the farmers market, the town manager or his designee, and selected gardeners. Working with these sponsors and customers, George develops the quality plan using information from the project charter and the project scope statement. The project charter contains project success criteria for vegetable produce quantities for the Community Gardens and individual gardens, and dollar revenue from farmers market sales. The project scope statement lists product characteristics, along with exclusions from the project and the product.

George takes the target metrics and allowable variances for the quality plan from the project charter and project scope statement. If target metric and variance information are not available, George consults with appropriate stakeholders and develops this information. He then updates the charter and scope statement as necessary and obtains stakeholder sign offs on the revised documents and the quality metrics table.

The quality plan for the Community Gardens is listed in Table I.1.

Table I.1 Quality Plan for Community Gardens Project

Priority [1 = high, 2 = medium, 3 = low]: 1-Scope; 2-Schedule; 3-Budget

PROJECT METRICS

CATEGORY	DESCRIPTION—PROJECT CAPABILITIES	TARGET METRIC	ALLOWABLE VARIANCE	ACTUAL PERFORMANCE
Scope	Project meets purpose as defined in charter	Contribute half of produce to soup kitchen Sell half of produce at farmers market Win 4H competition Gardeners report satisfaction with produce	None	
Schedule	Project meets schedule constraints	Schedule milestones as defined in project schedule	±10% or as defined in schedule	
Budget	Project meets budget constraints	Costs incurred as defined in project budget	±10% or as defined in budget	

(*Continued*)

Table I.1 (*Continued*) Quality Plan for Community Gardens Project

Priority [1 = high, 2 = medium, 3 = low]: 1-Scope; 2-Schedule; 3-Budget

PRODUCT METRICS

CATEGORY	DESCRIPTION—PRODUCT FEATURES	TARGET METRIC	ALLOWABLE VARIANCE	ACTUAL PERFORMANCE
Garden	Garden scope as specified in scope statement	Sq ft as specified in scope statement Fencing at specified height Sufficient faucets Enough mulch	±10% acceptable	
Produce	Produce quality equivalent to that found in local supermarkets	Not wilted, discolored, or bruised Fresh appearance No insects	±10% acceptable	
Support for community needy	Produce metrics	Generates 340 pounds of produce for soup kitchen Generates $3600 of farmers market sales	±10% acceptable	
Support for community in general	Project viewed as community asset	Survey of town's people report satisfaction with CG Survey of vendors, soup kitchen, farmers market say we are easy to work with	–30% (70% report satisfaction)	
Competition	4H competition	Entry accepted Entry wins	No acceptable variance	

Appendix J: Risks and Risk Categories

George has the risk categories from last year's Community Gardens project and from risk lists that he uses at work. He uses these as the initial risk categories for the Community Gardens project. George plans to modify this initial list as the project team identifies additional categories.

The list of risk categories is the same as the list used by the project team working the deck project. Table J.1 displays the risk categories.

Using the risk categories and working with the project stakeholders, the Community Gardens project team develops the following list of negative and positive risks. The risks below are in the form of a list; the list of risks for the deck was organized as a table with sections for project and product. Either format is fine. Different projects and different teams work best with different formats and structures.

J.1 Negative Risks (Threats)

- There is a risk that the Community Gardens will not have enough gardeners interested with the consequence that there won't be enough gardeners to work the individual plots and support the community section of the garden, and there won't be enough produce to support the soup kitchen and the farmers market.

Table J.1 Risk Categories

PRODUCT RISKS	ENVIRONMENTAL/EXTERNAL RISKS
• Requirements/Scope • Design & Technology • Quality & Feasibility	• Vendors • Legal/Regulatory/Safety • Market • Customer/Stakeholder
PROJECT ENVIRONMENTAL RISKS	PROJECT MANAGEMENT RISKS
• Decision making • Change management • Life cycle • Conflict management • Organization structure/Dysfunction	• Executive/Sponsor • Scope • Schedule • Budget/Costs • Quality • Resources

- There is a risk that the Community Gardens won't have volunteers to staff the weekly farmers market with a consequence that the project will not raise the targeted amount of money.
- There is a risk of pest infestation with a consequence that there will be less produce than planned.
- There is a risk that people will vandalize the gardens and take the produce with the consequence that there will not be enough produce for the soup kitchen and the farmers market.
- There is a risk that someone might get injured and sue the project or the town with a consequence of the cost and problem of legal action.
- There is a risk that the plants that purchased will be of poor quality or infected with pests with a consequence that the team will have to replace them, spending more than budgeted.

J.2 Positive Risks (Opportunities)

- There is a risk that the harvest will be greater than planned, with a result that the project will be able to provide additional produce to local food pantries.
- There is a risk that the project will have too many candidate gardeners with a consequence that the project manager will turn some gardeners away resulting in negative publicity for the project.

- There is an opportunity that the town paper publishes an article about the garden with the result that neighboring towns want the team's advice on setting up their own Community Gardens, taking time from the project manager and the individual gardeners.
- There is an opportunity that the quality and quantity of the produce will be so good that neighboring soup kitchens want donations, with a consequence that the project will be able to serve additional soup kitchens.
- There is an opportunity that the Community Gardens cookbook will be so excellent that the team is able to sell it on Amazon, generating far more revenue than planned with a consequence that the project is able to fund and support a second soup kitchen.

Sometimes an apparently positive risk has a negative consequence; the second and third positive risks in the list above have a negative consequence. The project manager doesn't eliminate the positive risk; he handles the situation such that the outcome is positive.

- Too many gardeners are an opportunity; turning away prospective gardeners may have a negative consequence. In this situation, the project manager focuses on a better alternative for dealing with the additional gardeners, such as obtaining an additional garden site or engaging these people to work the community portion of the gardens.
- A positive article in the town paper is good for the project. The project manager might delegate communications with the press and neighboring towns to one of the additional gardeners from above or might offer an internship to a local high school or community college student.

Appendix K: Risk Scores, Probability and Impact Scales

George has scales for probability and impact from last year's Community Gardens project. Since they worked well last year, George has the team review and validate them and then includes them in the project plan.

The Community Gardens project uses the same probability and impact scales as used for the deck project (Table K.1).

Applying these rankings to the negative and positive risks identified in Appendix J, the team develops risk scores and rankings.

For each risk, George leads the team in a discussion of an appropriate probability score (very high, high, low, or very low) and impact scores (high, medium, or low). He uses brainstorming and sticky note tactics to direct the discussion and decision-making tactics to drive the team discussion to a final decision. He records the probability and impact scores on the risk scores table and calculates a risk score; he ranks the risks based on the risk scores. The team also assigns each risk to a category. The quality plan included a relative importance of different project categories; the project manager uses the relative importance of the category to break a tie between risks. For example, if two risks have the same risk score, and one belongs to a priority

Table K.1 Probability and Impact Scales

PROBABILITY		IMPACT	
Very high	10	High	10
High	6	Medium	5
Low	4	Low	1
Very low	1		

category (such as scope, schedule, or budget), it takes precedence over the risk belonging to a lower priority category.

Based on the risk information in Table K.2, the risk rankings are as follows:

Negative risks starting with the most significant risks:

- Not enough produce due to vandalism—Risk Score = 60
- Not enough produce due to not enough gardeners—Risk Score = 40
- Not enough produce due to poor quality of plants or out-of-control garden pests—Risk score = 40
- Damage to fence or water supply (spigots, hoses) due to vandalism—Risk score = 40
- Reduced revenue due to not enough staff for farmers market—Risk Score = 6
- Increased expenses due to costs to settle claims as a result of injury—Risk Score = 5

Positive risks starting with the most significant risks:

- Because the quality/quantity of the produce is so good, the team partners with neighboring towns to generate extra produce and serve their soup kitchens—Risk score = 60.
- Able to recruit more gardeners and other volunteers because neighboring towns want the team's advice as a result of positive article in town paper—Risk Score = 50.
- The project serves additional food partners due to produce in excess of plan—Risk Score = 40.
- Excess revenue (due to selling more cookbooks than planned) allows the project to serve additional soup kitchens—Risk score = 40.

- There are too many gardeners so the project manager turns some away resulting in negative publicity; the opportunity is to obtain more land or create more, smaller plots for next year, generating more/better produce—Risk Score = 6.

Table K.2 Risk Scores for Community Gardens Project

NEGATIVE RISKS (THREATS)	CONSEQUENCE	RISK SCORE	PROB	IMPACT	CATEGORY
Not enough gardeners	Not enough produce	40	Low-4	High-10	Quality
Poor quality plants or pests	Not enough produce	40	Low-4	High-10	Quality
Vandalism	Not enough produce	60	High-6	High-10	Quality
	Damage to fence or water faucets	40	Low-4	High-10	Resource
Not enough staff for farmers market	Reduced revenue	6	High-6	Low-1	Budget
Injuries or liability	Cost to settle claim	5	Very Low-1	Med-5	Legal
POSITIVE RISKS (OPPORTUNITIES)	**CONSEQUENCE**	**RISK SCORE**	**PROB**	**IMPACT**	**CATEGORY**
Produce in excess of plan	Supply additional food partners	40	Low-4	High-10	Scope
Too many gardeners; turn away extras	Negative publicity; more land and/or plots for next year	6	High-6	Low-1	Market
Article in town paper	Neighboring towns want our advice; able to recruit more gardeners and other volunteers	50	Very high -10	Med-5	Schedule
Quality/Quantity of produce so good that neighboring towns want produce	Partner with neighboring towns to serve additional soup kitchens	60	High-6	High-10	Quality
Sell more cookbooks than planned	Excess revenue allows us to serve additional soup kitchens	40	Low-4	High-10	Budget

Appendix L: Risk Response Plans

George wants to maintain the project planning momentum and knows that continued stakeholder involvement depends on their enjoyment of, and involvement in, the process. He decides on a pot luck dinner and asks each stakeholder to bring a dish from the prior year's Community Gardens cookbook. He greets everyone at the door with a basket containing index cards; each index card contains one of the top-ranked risks from the list of risks for which the project team needs to develop a risk response plan. All the stakeholders assigned to the same risk (based on the content of the index cards) work together to develop a risk response plan. Stakeholders that have special subject matter expertise may switch to a different risk to ensure that the process uses stakeholder expertise appropriately.

At the end of the meeting, George collects the plans, circulates them to stakeholders for review and approval. He makes appropriate additions to the WBS, project schedule, and project budget and includes the risk plans and updates in the project plan.

For the Community Gardens, the most significant threat (negative risk) and opportunity (positive risk) are as follows:

- For threats, the top-rated risk is "there is a risk that people will break into the gardens and take produce with a consequence that the project will not have enough for the soup kitchen and the farmers market." The risk scores table (Table K.2) labeled this risk as "not enough produce due to vandalism" with a Risk Score = 60.
- For opportunities, the top-rated risk is "because the quality/ quantity of our produce is so good, the team partners with neighboring towns to generate extra produce and serve their soup kitchens" with a Risk Score = 60.

The negative and positive risk response plans are shown in Figure L.1, Risk Response Plan for CG Project—Negative Risk, and Figure L.2, Risk Response Plan for CG Project—Positive Risk.

Risk Response Plan

Risk Name: Theft of Produce **Date:**

Risk Category: Quality **Risk Owner:**

Risk Statement and Consequence: There is a risk that people will break into the gardens and take produce with a consequence that the team will not have enough for the soup kitchen and the farmers market.

Risk Score = 60 Probability: High–6 Impact: High–10

[For top ranked risks, complete the sections below.]

Response Strategies (Avoid/Exploit, Mitigate/Enhance, Transfer/Share, Accept (passive/active))

By choosing the avoid strategy, the team will take steps to prevent theft.

Actions to Implement the Response Strategies:

1. Keep Community Gardens locked when no one is around and working the gardens.
2. Post signs noting that unauthorized people are not permitted in the gardens, and that trespassers will be prosecuted. Post signs that tell people how they can benefit from the Community Gardens by going to the soup kitchen for meals.
3. Engage town police to monitor the gardens as part of their regular patrols.

Warning Signs of a Risk Occurrence

1. Fence and/or lock shows signs of tampering and/or are not secure.
2. Produce missing from gardens.
3. People loitering around the gardens when no one is working there.

Budget and Schedule

1. Work out patrol schedule with town police.
2. Add motion detectors, so intruders are startled and put in a spotlight. The plan is to borrow these from the town; if the team needs to purchase this equipment, there will be a budget impact that must be analyzed and approved by appropriate stakeholders.

Fallback Plans (What the project will do if risk occurs and primary response does not work.)

If police patrols, fencing, and signs do not deter theft of produce, the project manager might hire a night watchman to patrol for short periods of time. This will have a budget impact that must be analyzed and approved by appropriate stakeholders. Alternatively, the team may need to reassess the amount of produce they can supply to the soup kitchens. The key stakeholders will assess the viability of the Community Gardens in future years.

Residual Risks (Risks expected to remain after the risk response plan is implemented.)

If vandalism continues, there is risk that the stakeholders will not continue with the Community Gardens in future years.

Secondary Risks (Risks that occur as a consequence of implementing the risk response plan.)

As a consequence of securing the Community Gardens fence gates with locks, some gardeners might forget the lock combination or key and not have access to the gardens.

Contingency Reserves (Money set aside to cover the cost of the fallback plan.)

The fence and signs are provided by the town, so no additional cost is incurred by the project. The police patrols will be funded by the town. Potential additional costs include the cost of a night watchman. The project manager asks the sponsor to set aside $1000 as contingency. This money is outside the project budget; it is held by the sponsor in the town budget.

Figure L.1 Risk response plan for CG project—Negative risk.

Risk Response Plan

Risk Name: Excellent Quality and Quantity of Produce **Date:**

Risk Category: Quality **Risk Owner:**

Risk Statement and Consequence: Because the quality and quantity of our produce is so good, the team partners with neighboring towns to generate extra produce and serve their soup kitchens.

Risk Score = 60 Probability: High–6 Impact: High–10

[For top ranked risks, complete the sections below.]

Response Strategies (Avoid/Exploit, Mitigate/Enhance, Transfer/Share, Accept (passive/active))

By choosing the enhance strategy, the team will take steps to ensure excellent quality and quantity of produce:

1. The project manager enhances the probability of this risk by purchasing excellent plants from known vendors, and setting strict schedules for fertilizing, watering, and pest control.
2. The project has a plan to rotate crops from year to year to ensure that the soil is not depleted.

Actions to Implement the Response Strategies

1. Only purchase from plant vendors with a history of selling quality plants.
2. Publish schedules for fertilizing, watering, and pest control; monitor compliance via survey.
3. Educate interested folks in neighboring towns in our processes for growing, selling at the farmers market, and serving local soup kitchens.

Warning Signs of a Risk Occurrence

Positive comments from gardeners, soup kitchen manager, and farmers market customers

Budget and Schedule

No impact on schedule or budget

Fallback Plans (What the team will do if risk occurs and primary response does not work.)

1. None—if there are problems with plant quality and quantity, the team will accept this situation and revisit for future years.
2. No contingency plan if neighboring towns are not interested.

Residual Risks (Risks that the team expects will remain after the risk response plan is implemented.)

There is a risk that the neighboring towns might not be successful in implementing the processes that work well for Planful Gardens.

Secondary Risks (Risks that occur as a consequence of implementing the risk response plan.)

As neighboring towns begin to sell their produce at the farmers market, there may be a surplus of produce, driving prices down.

Contingency Reserves (Money set aside to cover the cost of the fallback plan.)

None—no cost to sharing processes with neighboring towns

Figure L.2 Risk response plan for CG project—Positive risk.

Appendix M: Stakeholder Matrix

George and the project team generate a list of stakeholders from project documents and in consultation with other stakeholders.

In the Community Gardens project, the stakeholders include:

- The project team, including the people who tend the gardens.
- Sponsors who fund the project (either money or resources). The town manager is a sponsor, in that the town provides resources for pre- and post-season garden preparation.
- Customers who consume the product of the project. In the Community Gardens, the soup kitchen, the farmers market, and the customers of the farmers market are all customers.
- The town newspaper might also be a stakeholder, in that it publicizes (either positively or negatively) the Community Gardens.

Stakeholders influence each other. Perhaps the project manager of the soup kitchen retires and another person takes on this work. Maybe the new project manager of the soup kitchen thinks that all community produce should go to the soup kitchen, and the project manager shouldn't sell produce at the farmers market. This new project manager of the soup kitchen is the wife of one of the vendors from whom the team purchases the garden plants. This vendor might decide that if the Community Gardens continue to sell produce at the farmers market, he will no longer be able to provide plants and materials at a reduced cost. While this is an unlikely (and political) scenario, it does point to the importance of continuing to manage and monitor the project stakeholders and their project bias.

The project team uses the power interest grid (Figure 10.1) to generate the relative priority of the project stakeholders. Priority is based on the power–interest scores. Priorities are as follows:

Priority 1 = *High Priority* based on high power and high interest
Priority 2 = *Medium High Priority* based on high power and low interest
Priority 3 = *Medium Low Priority* based on low power and high interest
Priority 4 = *Low Priority* based on low power and low interest

Based on the power interest scores on Table M.1, George ranks the relative importance of the stakeholders as follows:

Project manager—Priority = 1
Town manager—Priority = 1
Vendors of garden plants and supplies—Priority = 1
Town police—Priority = 2
Gardeners—Priority = 3
Soup kitchen manager and clients—Priority = 3
Farmers market lead—Priority = 3
Town maintenance workers—Priority = 3
Abutters—Priority = 3
Cookbook buyers—Priority = 4
Farmers market customers—Priority = 4

George reorders the stakeholder list based on the priority rank; the project team completes the stakeholder matrix for the stakeholders with the highest priority. If there are a great many stakeholders, the project team may complete the matrix columns 4 (what the stakeholder wants/needs from the project) and 5 (what the project needs from the stakeholder) only for the top-ranked stakeholders and keep the lower ranked stakeholders as a watch list. In the stakeholder matrix for the Community Gardens, George did not complete the matrix for stakeholders with a rank of 4 and for the soup kitchen clients (with a rank of 3). His decision to complete columns 4 and 5 for other stakeholders with a rank of 3 is based on their relatively more significant role in the project.

Table M.1 Stakeholder Matrix for Community Gardens Project

STAKEHOLDER NAME	POWER (P) INTEREST (I)	PRIORITY	WHAT STAKEHOLDER WANTS/NEEDS FROM THE PROJECT	WHAT THE PROJECT NEEDS FROM THE STAKEHOLDER
Project manager	High P High I	1	Complete project on scope, schedule, and budget	Strong leadership, direction, and communication
Gardeners	Low P High I	3	Garden plots ready to plant, customers for CG	Support and cultivation of the CG
Town manager (Sponsor)	High P High I	1	Good press, votes	Support in the form of town resources for garden prep, billing/accounting, and use of town vendors
Soup kitchen manager	Low P High I	3	Garden produce for meal preparation	Responsible use of the produce, no waste
Soup kitchen clients	Low P High I	3		
Farmers market lead	Low P High I	3	Reliable supply of good quality produce that can be sold	Consistent staffing of stand at market
Farmers market customers	Low P Low I	4		
Vendors-Garden plants & supplies	High P High I	1	Early notice of required materials and prompt payment	Top quality plants and supplies
Town police	High P Low I	2	No problems (no theft, vandalism, disorderly behavior)	Regular patrol and protection of gardens
Cookbook buyers	Low P Low I	4		
Town maintenance	Low P High I	3	Ease of preparing and maintaining the gardens	Timely preparation of garden plots
Abutters	Low P High I	3	No extra noise nor traffic	Congenial relations

Appendix N: Stakeholder Engagement Matrix

For the top-ranked stakeholders on the stakeholder matrix, George works with the project sponsor and other key stakeholders to complete the columns in the Stakeholder Engagement Matrix.

It is important that the top priority stakeholders are positively biased toward the project. Unaware and neutral are neither positive nor negative. Champions are positive stakeholders. From Table N.1, the top priority stakeholders (project manager, town manager, and vendors) are currently either neutral or positive, and they need to be positive or champions. This is feasible as it is a small move. Stakeholders that are negative (police, town maintenance, and abutters) are more challenging; George has to identify the WIIFM and convince them that the Community Gardens are good for them.

Since time is limited, George chooses to focus his attention on stakeholders with a priority of 1. The town manager is a top priority stakeholder with a neutral current engagement level. He is the project sponsor and it is important for the project to succeed that he be positive or a champion. From Table M.1, George knows that the town manager wants good press and votes from the project. George assesses the situation and develops a WIIFM to move the town manager from neutral to champion engagement level. The town manager

Table N.1 Stakeholder Engagement Matrix for Community Gardens Project

STAKEHOLDER NAME	POWER (P) INTEREST (I)	PRIORITY	UNAWARE	NEGATIVE	NEUTRAL	POSITIVE	CHAMPION
Project manager	High P High I	1				T	P
Gardeners	Low P High I	3			T	P	
Town manager	High P High I	1			T		P
Soup kitchen manager	Low P High I	3				T P	
Soup kitchen clients	Low P High I	3	T			P	
Farmers market lead	Low P High I	3				T P	
Farmers market customers	Low P Low I	4	T			P	
Vendors-Garden plants & supplies	High P High I	1			T	P	
Town police	High P Low I	2		T	P		
Cookbook buyers	Low P Low I	4	T		P		
Town maintenance	Low P High I	3		T	P		
Abutters	Low P High I	3		T	P		

Notes: P = Preferred Level of Engagement; T = Today's Level of Engagement.

may be newly elected and unaware of the Community Gardens. George explains how the Community Gardens enhances his popularity with the voters and suggests ways in which he can champion the project. Another situation might be if the town manager just heard about a lawsuit in a neighboring town, where a customer of the soup kitchen got sick from produce infected with bacteria and sued the town for millions of dollars. Now the town manager is negative and it will be difficult to move his level of engagement to positive. George identifies strategies to reassure the town manager. He explains the pesticide practices of the Community Gardens, including products used and frequency of application to reassure that the produce of the Community Gardens is safe. He also volunteers to change the Community Gardens' fertilizing practice to organic only.

George suggests that winning the 4H competition might enhance the town manager's reputation among competing towns, perhaps positioning him for election to a more powerful office in the future.

The vendors are currently neutral; perhaps notice of their generous discounts to the Community Gardens project in project marketing materials would promote more enthusiasm on the part of the vendors. George is currently positive; the sponsor (town manager) might note this and encourage a more leadership bias with public recognition of George's work on this important community project.

Stakeholders influence each other. The people whose property abuts the Community Gardens may object to the noise and traffic, especially in the evenings, generated by the Community Gardens, and they are negative stakeholders. George hears negative comments from the Community Gardens abutters at meetings and around town. Although the abutters are priority 3, George is concerned that they may negatively influence others in town. To mitigate this, George meets with the abutters to give them a chance to state their concerns. He assures them that activity at the Community Gardens won't continue past sunset as there is no lighting. He further assures them that he will post a sign that states that the gardens close at 8 p.m. and no people should work in the gardens after closing.

Appendix O: Project Communication Plan

For the top-ranked positive and negative stakeholders, George captures the project stakeholders' information requirements and documents necessary information in the project communication plan.

Projects often have many more stakeholders than the project manager and project team have time to address. George decides to complete the communication plan only for the top-ranked stakeholders. He copies the stakeholders and their priority rank from the stakeholder matrix (Table M.1). In the communication plan (Table O.1), George only provides additional information for stakeholders with a priority rank of 1 or 2. He leaves the lower ranked stakeholders on the list as a reminder to reach out to these stakeholders as time permits. He also completes the additional information for the gardeners (a priority 3) due to the large number of people in this stakeholder group.

George shares this information with the affected stakeholders and obtains their approvals.

Table O.1 Communication Plan for Community Gardens Project

RECIPIENT OF INFORMATION	CONTACT INFORMATION	INFORMATION NEEDED	TIME FRAME AND FREQUENCY	ACTION = SIGN OFF (S), REVIEW (R), EXECUTE (E)	WHO CREATES THE INFORMATION
Project manager Priority = 1	PM@gmail.com 999-888-7766	Status reports scope, time, cost	Fridays by 5 p.m.—weekly	R—reports	Soup kitchen manager; Farmers market lead
Gardeners Priority = 3	G@gmail.com 999-777-6655	Alert when plots ready			PM sends information
		Recruitment flyers; Kick off mtg notice; radio/TV spots	2 wks prior to kick off		PM sends information
Town manager (Sponsor) Priority = 1	TM@gmail.com 999-666-5544	Receipts, invoices,	Monday 5 p.m.—weekly	S—approve payments	PM generates receipts; Vendors generate invoices
		Good publicity	As it happens	S—approve external/ public messages	PM reports on publicity
Soup kitchen manager Priority = 3					
Soup kitchen clients Priority = 3					
Farmers market lead Priority = 3					
Farmers market customers Priority = 4					

(*Continued*)

Table 0.1 (*Continued*) Communication Plan for Community Gardens Project

RECIPIENT OF INFORMATION	CONTACT INFORMATION	INFORMATION NEEDED	TIME FRAME AND FREQUENCY	ACTION = SIGN OFF (S), REVIEW (R), EXECUTE (E)	WHO CREATES THE INFORMATION
Vendors of plants and supplies Priority = 1	V@yahoo.com	List of plants and supplies needed	Once per year by April 1	E—Order plants based on list	Gardeners generate list; PM aggregates all needs
Town police Priority = 2	Police@smalltown.com	Incident reports;	Incidents, as they happen	E—police respond to incidents	Community
		Patrol schedule	Monthly	E—Schedule	PM
Cookbook buyers Priority = 4					
Town workers Priority = 3					
Abutters Priority = 3					

Appendix P: Responsibility Assignment Matrices

George creates the responsibility assignment matrices (RAM) for project documents and product deliverables based on the project planning already complete. He has a list of project documents from the project plan, a list of deliverables from the WBS, and a list of stakeholders from the stakeholder matrix. He shares a draft of the RAMs with the appropriate stakeholders and obtains their sign off and approval before including them in the project plan.

Table P.1 lists the responsibilities for project documents for the Community Gardens project. Although George is responsible for most documents, he may choose to appoint an assistant to help him with this work; the project manager is still ultimately responsible.

Several stakeholders including soup kitchen clients, farmers market customers, cookbook buyers, and abutters to the Community Gardens have no responsibilities. These are all low priority stakeholders from the stakeholder engagement matrix. George lists them on the matrices so it is clear that they were not forgotten.

Table P.2 lists the responsibilities for WBS top level deliverables for the Community Gardens project.

Table P.1 RAM for Project Documents for Community Gardens Project

	PLANNING PHASE															EXECUTION		
	CHARTER	SCOPE STATEMENT	WBS	NETWORK DIAGRAM	SCHEDULE	BUDGET	LIST RISKS	RANKED RISKS	RISK RESP PLANS	STKHLDR MATRIX	ENGAGE MATRIX	COMM PLAN	RAM DOCS	RAM WBS	ORG CHART	ACTION ITEMLIST	STATUS REPORTS	CHANGE REQUESTS
Project manager	R	R	RS	RS	RS	R	RS	RS	RS	R	RS	RS	RS	RS	R	RS	R	R
Gardeners					C		C		C					I		I	I	I
Town manager	S	S				S				S				I	S		S	S
Soup kitchen manager					C		C		C					I			I	
Soup kitchen clients																		
Farmers market lead					C		C		C					I			I	
Farmers market customer																		
Vendor							C		C					I				
Police							C		C								I	
Cookbook buyers																		
Town maintenance							C		C									
Abutter																		

Notes: C—Stakeholder is a subject matter expert and responsible as a *consultant* or advisor; I—Stakeholder is *informed* of the document or activity but does not have to provide anything to the effort; R—Stakeholder is *responsible* for doing the activity or writing the document; S—Stakeholder is responsible for *sign off* on the document or activity.

Table P.2 RAM for WBS Deliverables for Community Gardens Project

	1.0– IND GARDEN PLANTED	2.0 PRODUCED HARVESTED	3.0 COMM PLOT PLANTED	4.0 PRE- & POST-SEASON PREP	5.0 COMM SERVICES PROVIDED	6.0 CG COOKBOOK SOLD	7.0 4H COMPETITION	8.0 PROJECT MANAGEMENT PROVIDED	9.0 PURCHASING COMPLETE
Project manager			S	S	S	R	S	R	R
Gardeners	RS	R	R		R		R		
Town manager (sponsor)								S	S
Soup kitchen manager		S			I				
Soup kitchen clients									
Farmers market lead		S			I				
Farmers market customer									
Vendor									C
Police									
Cookbook buyers						S			
Town maintenance				R					
Abutter									

Notes: C—Stakeholder is a subject matter expert and responsible as a *consultant* or advisor; I—Stakeholder is *informed* of the document or activity but does not have to provide anything to the effort; R—Stakeholder is *responsible* for doing the activity or writing the document; S—Stakeholder is responsible for *sign off* on the document or activity.

There are low priority stakeholders, including soup kitchen clients, farmers market customers, police, and abutters, who have no responsibilities for WBS deliverables.

George reviews this information with the appropriate stakeholders, obtains their approval, and includes the documents in the project plan.

Appendix Q: Organization Chart

George creates the organization chart based on how he expects the team reporting relationships to work. However, other stakeholders may have different expectations of the reporting relationships. For example, with the Community Gardens, the soup kitchen manager may view himself as a customer of the Community Gardens produce and therefore not reporting to George. George discusses this

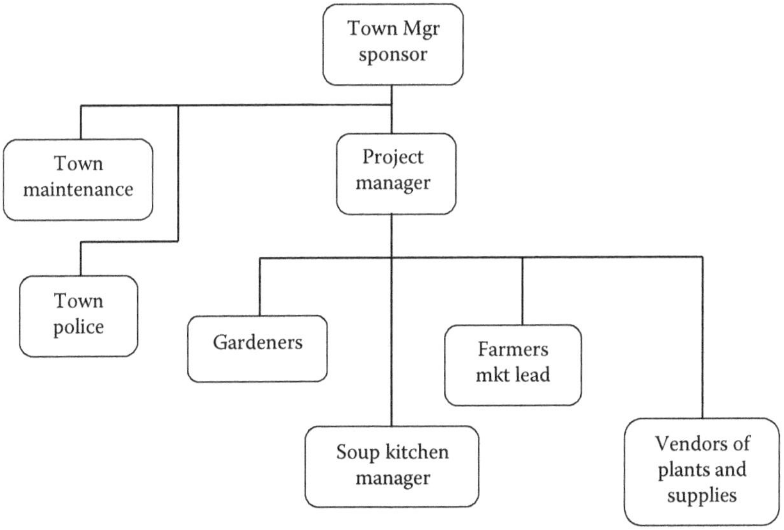

Figure Q.1 Organization chart for Community Gardens project.

with the soup kitchen manager, and he modifies the organization chart as necessary. Similarly, the vendor of plants and supplies may view himself as a vendor to the project and not part of the project organization chart. Again, this will be a topic for discussion, and the project organization chart may be modified as necessary (Figure Q.1).

Appendix R: Project Execution

At this point, the project managers have a good understanding of what the teams should be doing, but the teams have not yet begun to implement the project plan. George has a simple Community Gardens project and a project plan; Larry has very limited documentation. The two scenarios are Legume Legacy with Larry and Planful Gardens with George. Larry's project is significantly challenged by a lack of planning. George's project benefits from the development of a project plan.

Larry's team lacks a process. He missed the date for submission of budget items to the town; as a consequence, the team must fund the pre- and post-season garden plot preparation themselves. Larry doesn't have a list of prior years' gardeners so the team doesn't have enough resources. Since he doesn't have enough gardeners, he takes the excess land originally allocated to individual gardens and uses it to increase the overall size of the shared portion of the Community Gardens. As a result, he doesn't have enough gardeners to work the larger shared Community Gardens; the available land and untended produce is wasted. The team fails to deliver on its commitments to the soup kitchen and the farmers market. The 4H submission process leads to unresolved team conflict.

George's team has a well-documented process; the documents are listed in the appendices. George and his team complete the suggested documents and now have them available for reference and use in future years. Staffing is appropriate for the size of the Community Gardens and deliveries to the soup kitchen and farmers market are according to plan.

While Larry's team accomplished a lot for both the individual gardeners and also for the community services of the soup kitchen, farmers market, and 4H competition, they could have accomplished much more with a process and a project plan. George's team, with a documented process and plan, made better use of the town resources and the Community Gardens, returning more value, in the form of produce, to the soup kitchen, the farmers market, and the 4H competition.

R.1 The Scenario of George

As the Community Gardens' stakeholders get ready for the season kick off meeting, George looks at his communication plan. He notifies the volunteers from last year's list of the date and time for the kick off meeting via e-mail or telephone. Also, as documented in the communication plan, he posts a notice of the kick off meeting in the local newspaper and engages the local paper boys to post flyers about the meeting in the neighborhoods to interest new people in the Community Gardens project. He also has the local community radio and TV stations announce and advertise the Community Gardens project and the availability of garden plots.

George uses the kick off meeting agenda (Figure R.1) to assign responsibilities for the meeting; some people take meeting minutes, some people register the attendees, and other people arrange refreshments and entertainment.

George uses the agenda to run the meeting in an organized fashion. During the meeting, people sign up for individual plots, accept responsibility for maintenance of the community plots including a roster for planting, watering, and fertilizing. Gardeners sign up for the weeks that they will staff the farmers market and transport produce to the soup kitchen. George defers the responsibilities for the cookbook and 4H contest until later in the season.

George assigns a subcommittee to determine the crops for the Community Gardens and to collect the required materials and

Season Kick Off Meeting Agenda

April 21, 2015, 6–8 p.m.

Location: Town Hall Meeting Room, 12 Main St., Garden Town, MA

Meeting Lead—George, Project Manager

Note Taker—Sally Scribe

Register Attendees—Roger Register

Refreshments—Doug Dunkin

Entertainment—Sam Singer

6:00–6:15 p.m.: Introductions—everyone

6:15–7:00 p.m.: Garden plot assignments and responsibilities—George to lead

7:00–7:30 p.m.: Assignments to subcommittee for crops, materials, and supplies

7:30–8:00 p.m.: Discussion; Q&A; next steps

Figure R.1 Season kick off meeting agenda.

supplies for the individual gardens so that the team orders as a group and benefits from reduced prices. George monitors the subcommittee work to ensure that the plants are ordered in time for planting toward the end of May.

Once the season is underway, George monitors progress by MBWA—Managing By Walking Around. He visits the Community Gardens several times each week, on different days and at different times during the day. He talks with the gardeners to find out how things are going, and he monitors the status of the planting and the health of the plants. In this way, he has early warning of any problems and can take steps to fix these problems before they impact the goals of the Community Gardens. Talking with the gardeners, George discovers that they do not like using their personal cars to transport produce to the soup kitchen and the farmers market. One of the gardeners has a car that he is willing to loan to the Community Gardens if they pay the cost of the additional insurance. George's analysis of this potential change suggests that this is an excellent solution; the cost is small and funded from sales at the farmers market. George recommends approval of this change to the town manager (Figure R.2). Once approved, he updates the project budget, scope, and quality plan.

Change Request

Change Name: Loaner Car for Transportation of Produce and Staff **Date Required:**

Date Submitted: **Category:** Resource and Budget **Requested by:**

Description of Change: Gardeners do not like using their personal cars to transport produce to the soup kitchen and farmers market; gas is costly, and the vehicles are often needed by other family members. In addition, there is the liability with some near miss accidents. A town benefactor has an extra vehicle that he is willing to loan to the project if the Community Gardens pay the additional costs of insurance for additional drivers and liability coverage for use in a quasi-business. Additional coverage through the benefactor's policy is $500 for the remaining 3 months of deliveries to the soup kitchen and farmers market.

Benefits

1. Gardeners are happy not to use their own car; more reliable deliveries to the soup kitchen and farmers market.
2. Reduced liability for individual gardeners.

Limitations

Additional cost of $500

Impacts

Scope: additional resource in the form of a loaner car.

Schedule: no impact.

Budget: additional cost of $500 to be funded from receipts from produce sales at the farmers market.

Quality: more reliable deliveries to soup kitchen and farmers market.

Risks: no impact; liability risk is covered by additional insurance.

Recommendation

Recommend that the project team implement the change by accepting and using the loaner car. Request proof of insurance coverage from the car owner.

Reviewers and Date: ______________________

Approvers and Date: ______________________

Figure R.2 Change request and analysis—Loaner car.

As stated in the communication plan, George talks with the soup kitchen manager on a regular basis to ensure that he is getting the quantity and quality of produce promised. Sometimes George stops by the soup kitchen and other times he connects with the soup kitchen manager by e-mail or telephone. George employs a similar practice with the farmers market, stopping by periodically and communicating with the farmers market manager by e-mail or telephone. During this monitoring, George finds out that some of the soup kitchen deliveries are very late, and some of the folks are not fulfilling their assigned responsibilities to staff the farmers market stand. On discussion with the gardeners assigned to these tasks, George discovers that one lady has illness in the family and another has a scheduling conflict with the Thursday farmers market. George rearranges these staffing assignments to eliminate these conflicts.

In late July, George convenes a subcommittee to manage the 4H contest submission. They decide what produce to submit to the contest and complete the necessary contest paperwork. They also decide on staffing for the contest event and judging.

In October, George schedules the lessons learned and season-end event, as documented in the schedule. He begins the event with a short meeting, where the team creates lessons learned. They list what worked well and will do again in future years, what didn't work so well, and what they might do differently in the future. One suggestion is a list of volunteers, with contact information so that folks might arrange substitutes if they are unable to transport produce to the soup kitchen or staff the farmers market on their assigned day. George presents a summary of the Community Gardens' accomplishments, including pounds of produce delivered to the soup kitchen, dollar amount of funds raised from the farmers market, and status of the 4H contest (Figure R.3). He shares the post-project quality metrics for Community Gardens, as documented in Table R.1. Both the project and the product of the project were successful based on the comparison of post-project data with the target metrics.

Key stakeholders, invited to the year-end event, including the town manager, soup kitchen manager, farmers market manager, and 4H committee, are thrilled with the in-kind and financial contributions to the town's less fortunate people.

When the team finishes the business meeting, they bring out dishes for tasting and consideration for the season's Community Gardens cookbook. They fire up the grills and celebrate another successful Community Gardens season.

R.2 The Scenario of Larry

Larry has his own individual style of management for the Community Gardens. He doesn't have a list of gardeners from last year; he contacts those gardeners from last year for whom he has telephone numbers or e-mail addresses. He talks up the Community Gardens with his colleagues at the golf course and the local coffee shop, but he only has gardeners for about half of the individual plots. Since the individual

Lessons Learned

Summary of Community Garden Project

1. Pounds of product delivered to soup kitchen = approximately 350 pounds; in line with expectations in project charter.
2. Funds raised from farmers market sales = $3400; just below objectives in project charter.
3. 4H contest = first place winner.

What Worked Well

1. Gardeners worked well together; got along well as a team.
2. The project plan guided the planting and harvesting process.

What Did Not Work Well

1. Subcommittee planning for 4H contest submissions needed more time; start this planning earlier in the future.
2. It was difficult to transport produce in the gardeners' personal vehicles; try to get use of town truck in future years.

Suggestions

1. Create and distribute list of gardeners and volunteers so if someone cannot fulfill his responsibility for soup kitchen deliveries or staffing of the farmers market, he knows whom to call and how to contact substitute volunteers.
2. Get town truck to transport produce in future years.

Figure R.3 Lessons learned—Community Gardens project.

gardeners also staff the shared plot for soup kitchen and farmers market produce, there are fewer gardeners to share this work also.

Larry holds the kick off meeting and generates a lot of enthusiasm for the project. He has people sign up for multiple individual garden plots with the understanding that, as more gardeners sign up for the project, these garden plots will be reassigned to these newcomers. Lacking a schedule, Larry forgets to submit paperwork to the town, and the gardeners must pay for pre- and post-season ground preparation on their own. He doesn't aggregate the equipment and material needs from individual gardeners and the community plot so he is unable to take advantage of discounted pricing from town approved vendors. Eventually, the gardens are planted and even though the Community Gardens lack a schedule of responsibility for watering and fertilizing, the produce in the community garden seems to be thriving.

Table R.1 Quality Plan for Community Gardens Project—Post-Project Completion

Priority [1 = high, 2 = medium, 3 = low]: 1-Scope; 2-Schedule; 3-Budget

PROJECT METRICS

CATEGORY	DESCRIPTION—PROJECT CAPABILITIES	TARGET METRIC	ALLOWABLE VARIANCE	ACTUAL PERFORMANCE
Scope	Project meets purpose as defined in charter	Contribute half of produce to soup kitchen Sell half of produce at farmers market Win 4H competition Gardeners report satisfaction with produce	None	Met target metric
Schedule	Project meets schedule constraints	Schedule milestones as defined in project schedule	±10% or as defined in schedule	Occasional delays in deliveries to soup kitchen
Budget	Project meets budget constraints	Costs incurred as defined in project budget	±10% or as defined in budget	Within target for modified budget of $6000

(*Continued*)

Table R.1 (*Continued*) Quality Plan for Community Gardens Project—Post-Project Completion

Priority [1 = high, 2 = medium, 3 = low]: 1-Scope; 2-Schedule; 3-Budget

PRODUCT METRICS

CATEGORY	DESCRIPTION—PRODUCT FEATURES	TARGET METRIC	ALLOWABLE VARIANCE	ACTUAL PERFORMANCE
Garden	Garden scope as specified in scope statement	Sq ft as specified in scope statement Fencing at specified height Sufficient faucets Enough mulch	±10% acceptable	Met target metric Gardeners report faucets not in convenient locations
Produce	Produce quality equivalent to that found in local supermarkets	Not wilted, discolored, or bruised Fresh appearance No insects	±10% of produce meeting metric acceptable	Met target spec
Support for community needy	Produce metrics	Generates 340 pounds of produce for soup kitchen Generates $3600 of farmers market sales	±10% acceptable	Met target spec
Support for community in general	Project viewed as community asset	Survey of town's people report satisfaction with CG Survey of vendors, soup kitchen, farmers market say we are easy to work with	–30% (70% report satisfaction)	Exceeded target by 5%
Competition	4H competition	Entry accepted Entry wins	No acceptable variance	Met target spec

Larry doesn't monitor the soup kitchen and the farmers market—he believes in a hands-off approach and is adverse to micromanaging. As a result, he doesn't find out that deliveries to the soup kitchen have been sporadic and the farmers market stand was staffed only half the time. Actually, the soup kitchen manager called several times, but Larry didn't quite understand the message and didn't follow up to clarify. This was a very big issue from the perspective of the soup kitchen manager. Unknown to Larry, the soup kitchen has a federal grant that matched all contributions from local sources. The lack of planned produce from the Community Gardens resulted in a reduction in the monies received from the grant with a consequence that the soup kitchen was not able to feed the expected number of people. It is important to understand that project commitments often have implications that go far beyond the project scope.

Larry's team missed the deadline for submitting paperwork to the 4H sponsored contest, so the team passed on the event this year. At the year-end event, Larry skipped the lessons learned and the team went directly to celebrating with good food and drink.

R.3 Conclusion

Did the existence of a process and a plan make a difference? Larry's team did contribute to the less fortunate in the town and used the Community Gardens project to generate a sense of community pride. But, given the resources available, they could have returned a great deal more to the community, the soup kitchen, and the town's less fortunate. The lack of a process also precluded a more effective effort in the following year. Lacking a process and documentation on what worked and what didn't work, Larry's team is likely to repeat the same mistakes and incur the same limitations in future years.

George and the Planful Garden team developed a project plan and implemented this plan to guide the development of the Community Gardens deliverables, including individual gardens, community garden, soup kitchen, farmers market, 4H competition, and cookbook. The Community Gardens project is predictable and well managed.

Index

www.ingramcontent.com/pod-product-compliance
Lightning Source LLC
LaVergne TN
LVHW011119300926
849160LV00008B/445
9781498729345